Solutions Manual
to Accompany

Accelerated Studies in Physics and Chemistry

Rebekah L. Mays and John D. Mays

NOVARE
SCIENCE & MATH

Austin, Texas
2015

Published by

Novare Science & Math LLC
P. O. Box 92934
Austin, Texas 78709-2934
novarescienceandmath.com

NOVARE
SCIENCE & MATH

Printed in the United States of America

ISBN: 978-0-9966771-6-5

For the complete catalog of textbooks and resources available from Novare Science & Math, visit novarescienceandmath.com.

Contents

Acknowledgement

I (John) wish to express my gratitude to my daughter Rebekah Mays for carefully and meticulously compiling these solutions. It may have been a walk down memory lane for her: she took this course back in 04–05 when she was in 9th grade. (I was the teacher.)

Any errors that remain in this volume are my own responsibility.

Preface

This solutions manual contains fully detailed solutions for all of the computational problems contained in my text *Accelerated Studies in Physics and Chemistry*. Teachers and students using that text should find this manual to be a valuable resource.

When comparing your results to the results shown here and to those in the text, keep in mind that the last digit is always uncertain because of the way significant digits in measurements are defined. When two results match except for a small difference in the most precise digit, we say that the results match. Because of rounding in calculators, it will not be uncommon for results shown here to differ from the answer key in the text or from your result by or two in the most precise digit.

I have checked and double checked the solutions to make them as accurate as possible. However, in any manual of this kind it is inevitable that errors remain. If you find an error, we would be much obliged if you would inform us of it by sending an email to info@novarescienceandmath.com.

Chapter 2

Unit Conversions

1.

$$1{,}750 \text{ m} \cdot \frac{100 \text{ cm}}{1 \text{ m}} \cdot \frac{1 \text{ in}}{2.54 \text{ cm}} \cdot \frac{1 \text{ ft}}{12 \text{ in}} = 5{,}740 \text{ ft}$$

2.

$$3.54 \text{ g} \cdot \frac{1 \text{ kg}}{1000 \text{ g}} = 0.00354 \text{ kg}$$

3.

$$41.11 \text{ mL} \cdot \frac{1 \text{ L}}{1000 \text{ mL}} = 0.04111 \text{ L}$$

4.

$$7 \times 10^{8} \text{ m} \cdot \frac{100 \text{ cm}}{1 \text{ m}} \cdot \frac{1 \text{ in}}{2.54 \text{ cm}} \cdot \frac{1 \text{ ft}}{12 \text{ in}} \cdot \frac{1 \text{ mi}}{5{,}280 \text{ ft}} = 4 \times 10^{5} \text{ mi}$$

5.

$$1.5499 \times 10^{-12} \text{ mm} \cdot \frac{1 \text{ m}}{1000 \text{ mm}} = 1.5499 \times 10^{-15} \text{ m}$$

6.

$$750 \text{ cm}^{3} \cdot \frac{1 \text{ mL}}{1 \text{ cm}^{3}} \cdot \frac{1 \text{ L}}{1000 \text{ mL}} \cdot \frac{1 \text{ m}^{3}}{1000 \text{ L}} = 7.5 \times 10^{-4} \text{ m}^{3}$$

7.

$$2.9979 \times 10^{8} \frac{\text{m}}{\text{s}} \cdot \frac{100 \text{ cm}}{1 \text{ m}} \cdot \frac{1 \text{ in}}{2.54 \text{ cm}} \cdot \frac{1 \text{ ft}}{12 \text{ in}} = 9.836 \times 10^{8} \frac{\text{ft}}{\text{s}}$$

8.

$$168 \text{ hr} \cdot \frac{60 \text{ min}}{1 \text{ hr}} \cdot \frac{60 \text{ s}}{1 \text{ min}} = 605{,}000 \text{ s}$$

9.

$$5{,}570 \frac{\text{kg}}{\text{m}^{3}} \cdot \frac{1000 \text{ g}}{1 \text{ kg}} \cdot \frac{1 \text{ m}^{3}}{1000 \text{ L}} \cdot \frac{1 \text{ L}}{1000 \text{ mL}} \cdot \frac{1 \text{ mL}}{1 \text{ cm}^{3}} = 5.57 \frac{\text{g}}{\text{cm}^{3}}$$

10.

$$45 \frac{\text{gal}}{\text{s}} \cdot \frac{3.786 \text{ L}}{1 \text{ gal}} \cdot \frac{1 \text{ m}^3}{1000 \text{ L}} \cdot \frac{60 \text{ s}}{1 \text{ min}} = 1.0 \times 10^1 \frac{\text{m}^3}{\text{min}}$$

11.

$$600,000 \frac{\text{ft}^3}{\text{s}} \cdot \frac{(0.3048 \text{ m})^3}{1 \text{ ft}^3} \cdot \frac{1000 \text{ L}}{1 \text{ m}^3} \cdot \frac{60 \text{ s}}{1 \text{ min}} \cdot \frac{60 \text{ min}}{1 \text{ hr}} = 6 \times 10^{10} \frac{\text{L}}{\text{hr}}$$

12.

$$5,200 \text{ mL} \cdot \frac{1 \text{ L}}{1000 \text{ mL}} \cdot \frac{1 \text{ m}^3}{1000 \text{ L}} = 5.2 \times 10^{-3} \text{ m}^3$$

13.

$$5.65 \times 10^2 \text{ mm}^2 \cdot \frac{1 \text{ cm}}{10 \text{ mm}} \cdot \frac{1 \text{ cm}}{10 \text{ mm}} \cdot \frac{1 \text{ in}}{2.54 \text{ cm}} \cdot \frac{1 \text{ in}}{2.54 \text{ cm}} = 0.876 \text{ in}^2$$

14.

$$32.16 \frac{\text{ft}}{\text{s}^2} \cdot \frac{12 \text{ in}}{1 \text{ ft}} \cdot \frac{2.54 \text{ cm}}{1 \text{ in}} \cdot \frac{1 \text{ m}}{100 \text{ cm}} = 9.802 \frac{\text{m}}{\text{s}^2}$$

15.

$$5.001 \frac{\mu \text{g}}{\text{s}} \cdot \frac{1 \text{ g}}{10^6 \text{ } \mu \text{g}} \cdot \frac{1 \text{ kg}}{1000 \text{ g}} \cdot \frac{60 \text{ s}}{1 \text{ min}} = 3.001 \times 10^{-4} \frac{\text{kg}}{\text{min}}$$

16.

$$4.771 \frac{\text{g}}{\text{mL}} \cdot \frac{1 \text{ kg}}{1000 \text{ g}} \cdot \frac{1000 \text{ mL}}{1 \text{ L}} \cdot \frac{1000 \text{ L}}{1 \text{ m}^3} = 4,771 \frac{\text{kg}}{\text{m}^3}$$

17.

$$13.6 \frac{\text{g}}{\text{cm}^3} \cdot \frac{1000 \text{ mg}}{1 \text{ g}} \cdot \frac{100 \text{ cm}}{1 \text{ m}} \cdot \frac{100 \text{ cm}}{1 \text{ m}} \cdot \frac{100 \text{ cm}}{1 \text{ m}} = 1.36 \times 10^{10} \frac{\text{mg}}{\text{m}^3}$$

18.

$$93,000,000 \text{ mi} \cdot \frac{5280 \text{ ft}}{1 \text{ mi}} \cdot \frac{0.3048 \text{ m}}{1 \text{ ft}} \cdot \frac{100 \text{ cm}}{1 \text{ m}} = 1.5 \times 10^{13} \text{ cm}$$

19.

$$65 \frac{\text{mi}}{\text{hr}} \cdot \frac{5,280 \text{ ft}}{1 \text{ mi}} \cdot \frac{0.3048 \text{ m}}{1 \text{ ft}} \cdot \frac{1 \text{ hr}}{60 \text{ min}} \cdot \frac{1 \text{ min}}{60 \text{ s}} = 29 \frac{\text{m}}{\text{s}}$$

20.

$$633 \text{ nm} \cdot \frac{1 \text{ m}}{10^9 \text{ nm}} \cdot \frac{100 \text{ cm}}{1 \text{ m}} \cdot \frac{1 \text{ in}}{2.54 \text{ cm}} = 2.49 \times 10^{-5} \text{ in}$$

21.

$$0.05015 \cdot 3.00 \times 10^8 \frac{\text{m}}{\text{s}} \cdot \frac{60 \text{ s}}{1 \text{ min}} \cdot \frac{60 \text{ min}}{1 \text{ hr}} \cdot \frac{1 \text{ ft}}{0.3048 \text{ m}} \cdot \frac{1 \text{ mi}}{5,280 \text{ ft}} = 3.37 \times 10^7 \frac{\text{mi}}{\text{hr}}$$

22.

$T_F = 98.6°F$

$T_C = ?$

$$T_C = \frac{5}{9}(T_F - 32) = \frac{5}{9}(98.6°F - 32) = 37.0°C$$

23.

$T_C = 50.0°C$

$T_F = ?$

$$T_C = \frac{5}{9}(T_F - 32)$$

$$T_F = \frac{9}{5}T_C + 32 = \frac{9}{5}(50.0°C) + 32 = 122°F$$

24.

$$1 \text{ lt-yr} = 3.00 \times 10^8 \frac{\text{m}}{\text{s}} \cdot \frac{60 \text{ s}}{1 \text{ min}} \cdot \frac{60 \text{ min}}{1 \text{ hr}} \cdot \frac{24 \text{ hr}}{1 \text{ day}} \cdot \frac{365 \text{ days}}{1 \text{ year}} = 9.461 \times 10^{15} \text{ m}$$

$$4.3 \text{ lt-yr} \cdot 9.461 \times 10^{15} \frac{\text{m}}{\text{lt-yr}} \cdot \frac{1 \text{ km}}{1000 \text{ m}} = 4.1 \times 10^{13} \text{ km}$$

Motion Study Questions Set 1

1.

$$d = 25.1 \text{ mi} \cdot \frac{5,280 \text{ ft}}{1 \text{ mi}} \cdot \frac{0.3048 \text{ m}}{1 \text{ ft}} = 4.04 \times 10^4 \text{ m}$$

$$t = 0.50 \text{ hr} \cdot \frac{60 \text{ min}}{1 \text{ hr}} \cdot \frac{60 \text{ s}}{1 \text{ min}} = 1,800 \text{ s}$$

$$v = ?$$

$$d = vt$$

$$v = \frac{d}{t} = \frac{4.04 \times 10^4 \text{ m}}{1,800 \text{ s}} = 22 \frac{\text{m}}{\text{s}}$$

2.

$$22 \frac{\text{m}}{\text{s}} \cdot \frac{1 \text{ km}}{1000 \text{ m}} \cdot \frac{60 \text{ s}}{1 \text{ min}} \cdot \frac{60 \text{ min}}{1 \text{ hr}} = 79 \frac{\text{km}}{\text{hr}}$$

3.

$$t = 4.25 \text{ hr}$$

$$v = 5.0000 \frac{\text{km}}{\text{hr}}$$

$$d = ?$$

$$d = vt$$

$$d = 5.0000 \frac{\text{km}}{\text{hr}} \cdot 4.25 \text{ hr} = 21.3 \text{ km}$$

4.

$$21.3 \text{ km} \cdot \frac{1000 \text{ m}}{1 \text{ km}} \cdot \frac{1 \text{ ft}}{0.3048 \text{ m}} \cdot \frac{1 \text{ mi}}{5,280 \text{ ft}} = 13.2 \text{ mi}$$

5.

$$150.0 \frac{\text{mi}}{\text{hr}} \cdot \frac{5,280 \text{ ft}}{1 \text{ mi}} \cdot \frac{0.3048 \text{ m}}{1 \text{ ft}} \cdot \frac{1 \text{ km}}{1000 \text{ m}} = 241.4 \frac{\text{km}}{\text{hr}}$$

6.

$$v = 150.0 \ \frac{\text{mi}}{\text{hr}} \cdot \frac{1 \text{ hr}}{60 \text{ min}} = 2.50 \ \frac{\text{mi}}{\text{min}}$$

$d = 10.0 \text{ mi}$

$t = ?$

$d = vt$

$$t = \frac{d}{v}$$

$$t = \frac{10.0 \text{ mi}}{2.50 \ \frac{\text{mi}}{\text{min}}} = 4.00 \text{ min}$$

7.

$$d = 3.0 \text{ km} \cdot \frac{1000 \text{ m}}{1 \text{ km}} = 3.0 \times 10^3 \text{ m}$$

$$t = 1 \text{ hr } 20.0 \text{ min} = 80.0 \text{ min} \cdot \frac{60 \text{ s}}{1 \text{ min}} = 4.80 \times 10^3 \text{ s}$$

$v = ?$

$d = vt$

$$v = \frac{d}{t} = \frac{3.0 \times 10^3 \text{ m}}{4.80 \times 10^3 \text{ s}} = 0.63 \ \frac{\text{m}}{\text{s}}$$

8.

$v_i = 0$

$$v_f = 45 \ \frac{\text{mi}}{\text{hr}} \cdot \frac{1 \text{ hr}}{60 \text{ min}} \cdot \frac{1 \text{ min}}{60 \text{ s}} \cdot \frac{5,280 \text{ ft}}{1 \text{ mi}} \cdot \frac{0.3048 \text{ m}}{1 \text{ ft}} = 20.1 \ \frac{\text{m}}{\text{s}}$$

$t = 36 \text{ s}$

$a = ?$

$$a = \frac{v_f - v_i}{t} = \frac{20.1 \frac{\text{m}}{\text{s}} - 0}{36 \text{ s}} = 0.56 \ \frac{\text{m}}{\text{s}^2}$$

9.

$$v_i = 31 \ \frac{m}{s}$$

$$t = 17 \ s$$

$$v_f = 22 \ \frac{m}{s}$$

$$a = ?$$

$$a = \frac{v_f - v_i}{t} = \frac{22 \ \frac{m}{s} - 31 \ \frac{m}{s}}{17 \ s} = -0.53 \ \frac{m}{s^2}$$

10.

$$d = 14.5 \ m$$

$$v = c = 3.00 \times 10^8 \ \frac{m}{s}$$

$$t = ?$$

$$d = vt$$

$$t = \frac{d}{v} = \frac{14.5 \ m}{3.00 \times 10^8 \ \frac{m}{s}} = 4.83 \times 10^{-8} \ s \cdot \frac{10^9 \ ns}{s} \ 48.3 \ ns$$

11.

$$v_i = 0$$

$$v_f = 0.80 \cdot 3.00 \times 10^8 \ \frac{m}{s} = 2.40 \times 10^8 \ \frac{m}{s}$$

$$t = 18 \ hr \ 6 \ min \ 45 \ s = 64800 \ s + 360 \ s + 45 \ s = 65,205 \ s$$

$$a = ?$$

$$a = \frac{v_f - v_i}{t} = \frac{2.40 \times 10^8 \ \frac{m}{s} - 0}{65,205 \ s} = 3,680 \ \frac{m}{s^2}$$

12.

$$d = 8.96 \times 10^9 \text{ km} \cdot \frac{1000 \text{ m}}{1 \text{ km}} = 8.96 \times 10^{12} \text{ m}$$

$$v = 3.45 \times 10^5 \ \frac{\text{m}}{\text{s}}$$

$$t = ?$$

$$d = vt$$

$$t = \frac{d}{v} = \frac{8.96 \times 10^{12} \text{ m}}{3.45 \times 10^5 \ \frac{\text{m}}{\text{s}}} = 2.597 \times 10^7 \text{ s} \cdot \frac{1 \text{ min}}{60 \text{ s}} \cdot \frac{1 \text{ hr}}{60 \text{ min}} \cdot \frac{1 \text{ day}}{24 \text{ hr}} = 301 \text{ days}$$

13.

$$a = 5.556 \times 10^6 \ \frac{\text{cm}}{\text{s}^2} \cdot \frac{1 \text{ m}}{100 \text{ cm}} = 5.556 \times 10^4 \ \frac{\text{m}}{\text{s}^2}$$

$$t = 45 \text{ ms} \cdot \frac{1 \text{ s}}{1000 \text{ ms}} = 4.5 \times 10^{-2} \text{ s}$$

$$v_i = 0$$

$$v_f = ?$$

$$a = \frac{v_f - v_i}{t}$$

$$v_f = at + v_i = (5.556 \times 10^4 \ \frac{\text{m}}{\text{s}^2})(4.5 \times 10^{-2} \text{ s}) + (0 \ \frac{\text{m}}{\text{s}}) = 2.5 \times 10^3 \ \frac{\text{m}}{\text{s}}$$

14.

$$v_i = 4.005 \times 10^3 \ \frac{\text{m}}{\text{s}}$$

$$a = 23.1 \ \frac{\text{m}}{\text{s}^2}$$

$$t = 13.5 \text{ s}$$

$$v_f = ?$$

$$a = \frac{v_f - v_i}{t}$$

$$v_f = at + v_i = (23.1 \ \frac{\text{m}}{\text{s}^2} \cdot 13.5 \text{ s}) + 4.005 \times 10^3 \ \frac{\text{m}}{\text{s}} = 4.32 \times 10^3 \ \frac{\text{m}}{\text{s}}$$

15.

$$v = c = 2.9979 \times 10^8 \ \frac{\text{m}}{\text{s}}$$

$$d = 1.4965 \times 10^8 \ \text{km} \cdot \frac{1000 \ \text{m}}{1 \ \text{km}} = 1.4965 \times 10^{11} \ \text{m}$$

$$t = ?$$

$$d = vt$$

$$t = \frac{d}{v} = \frac{1.4965 \times 10^{11} \ \text{m}}{2.9979 \times 10^8 \ \frac{\text{m}}{\text{s}}} = 499.18 \ \text{s} \cdot \frac{1 \ \text{min}}{60 \ \text{s}} = 8.3197 \ \text{min}$$

Chapter 3

Newton's Second Law Practice Problems

1.

$m = 1{,}880$ kg

$a = 1.50 \ \dfrac{m}{s^2}$

$F = ?$

$a = \dfrac{F}{m}$

$F = ma = 1{,}880 \text{ kg} \cdot 1.50 \ \dfrac{m}{s^2} = 2{,}820$ N

2.

$m = 188.4 \text{ g} \cdot \dfrac{1 \text{ kg}}{1000 \text{ g}} = 0.1884$ kg

$g = 9.80 \ \dfrac{m}{s^2}$

$F_w = ?$

$F_w = 0.1884 \text{ kg} \cdot 9.80 \ \dfrac{m}{s^2} = 1.85$ N

3.

$F = 250.0$ N

$m = 144{,}000 \text{ mg} \cdot \dfrac{1 \text{ g}}{1000 \text{ mg}} \cdot \dfrac{1 \text{ kg}}{1000 \text{ g}} = 0.144$ kg

$a = ?$

$a = \dfrac{F}{m} = \dfrac{250.0 \text{ N}}{0.144 \text{ kg}} = 1{,}740 \ \dfrac{m}{s^2}$

4.

$$a = 2.3 \, \frac{m}{s^2}$$

$$F = 230,000 \, N$$

$$m = ?$$

$$a = \frac{F}{m}$$

$$m = \frac{F}{a} = \frac{230,000 \, N}{2.3 \, \frac{m}{s^2}} = 1.0 \times 10^5 \, kg$$

5.

$$a = 0.0022 \, \frac{mi}{hr^2} \cdot \frac{5,280 \, ft}{1 \, mi} \cdot \frac{0.3048 \, m}{1 \, ft} \cdot \frac{1 \, hr}{60 \, min} \cdot \frac{1 \, hr}{60 \, min} \cdot \frac{1 \, min}{60 \, s} \cdot \frac{1 \, min}{60 \, s} = 2.732 \times 10^{-7} \, \frac{m}{s^2}$$

$$m = 2.2 \, Mg \cdot \frac{10^6 \, g}{1 \, Mg} \cdot \frac{1 \, kg}{1000 \, g} = 2.2 \times 10^3 \, kg$$

$$F = ?$$

$$a = \frac{F}{m}$$

$$F = ma = 2.2 \times 10^3 \, kg \cdot 2.732 \times 10^{-7} \, \frac{m}{s^2} = 6.0 \times 10^{-4} \, N$$

6.

$$F_w = 125.1 \, lb \cdot \frac{4.45 \, N}{1 \, lb} = 556.7 \, N$$

$$g = 9.80 \, \frac{m}{s^2}$$

$$m = ?$$

$$F_w = mg$$

$$m = \frac{F_w}{g} = \frac{556.7 \, N}{9.80 \, \frac{m}{s^2}} = 56.8 \, kg$$

7.

$m = 56.8$ kg

$F_w = 17.9$ lb $\cdot \dfrac{4.45 \text{ N}}{1 \text{ lb}} = 79.66$ N

$g_m = ?$

$F_w = mg_m$

$g_m = \dfrac{F_w}{m} = \dfrac{79.66 \text{ N}}{56.8 \text{ kg}} = 1.40 \ \dfrac{\text{m}}{\text{s}^2}$

8.

$v_i = 0$

$v_f = 125.0 \ \dfrac{\text{m}}{\text{s}}$

$t = 22.00$ ms $\cdot \dfrac{1 \text{ s}}{1000 \text{ ms}} = 2.200 \times 10^{-2}$ s

$F = 142.0$ N

$m = ?$

$a = \dfrac{v_f - v_i}{t} = \dfrac{125.0 \ \dfrac{\text{m}}{\text{s}} - 0}{2.200 \times 10^{-2} \text{ s}} = 5{,}681.8 \ \dfrac{\text{m}}{\text{s}^2}$

$a = \dfrac{F}{m}$

$m = \dfrac{F}{a} = \dfrac{142.0 \text{ N}}{5681.8 \ \dfrac{\text{m}}{\text{s}^2}} = 0.024992 \text{ kg} \cdot \dfrac{1000 \text{ g}}{1 \text{ kg}} = 24.99$ g

---

9.

$m = 4.5 \text{ kg}$

$v_i = 0$

$v_f = 8.00 \dfrac{\text{mi}}{\text{hr}} \cdot \dfrac{1 \text{ hr}}{60 \text{ min}} \cdot \dfrac{1 \text{ min}}{60 \text{ s}} \cdot \dfrac{5{,}280 \text{ ft}}{1 \text{ mi}} \cdot \dfrac{0.3048 \text{ m}}{1 \text{ ft}} = 3.576 \dfrac{\text{m}}{\text{s}}$

$t = 500 \text{ ms} \cdot \dfrac{1 \text{ s}}{1000 \text{ ms}} = 0.5 \text{ s}$

$F = ?$

$a = \dfrac{v_f - v_i}{t} = \dfrac{3.576 \dfrac{\text{m}}{\text{s}} - 0}{0.5 \text{ s}} = 7.152 \dfrac{\text{m}}{\text{s}^2}$

$a = \dfrac{F}{m}$

$F = ma = 4.5 \text{ kg} \cdot 7.152 \dfrac{\text{m}}{\text{s}^2} = 30 \text{ N}$

10.

$v_i = 2{,}500.0 \dfrac{\text{km}}{\text{hr}} \cdot \dfrac{1 \text{ hr}}{60 \text{ min}} \cdot \dfrac{1 \text{ min}}{60 \text{ s}} \cdot \dfrac{1000 \text{ m}}{1 \text{ km}} = 694.4 \dfrac{\text{m}}{\text{s}}$

$t = 8.000 \text{ s}$

$F = 45{,}450 \text{ N}$

$v_f = 2{,}750 \dfrac{\text{km}}{\text{hr}} \cdot \dfrac{1 \text{ hr}}{60 \text{ min}} \cdot \dfrac{1 \text{ min}}{60 \text{ s}} \cdot \dfrac{1000 \text{ m}}{1 \text{ km}} = 763.9 \dfrac{\text{m}}{\text{s}}$

$m = ?$

$a = \dfrac{v_f - v_i}{t} = \dfrac{763.89 \dfrac{\text{m}}{\text{s}} - 694.44 \dfrac{\text{m}}{\text{s}}}{8.000 \text{ s}} = 8.681 \dfrac{\text{m}}{\text{s}^2}$

$a = \dfrac{F}{m}$

$m = \dfrac{F}{a} = \dfrac{45{,}450 \text{ N}}{8.681 \dfrac{\text{m}}{\text{s}^2}} = 5{,}230 \text{ kg}$

11.

$$m = 166 \text{ g} \cdot \frac{1 \text{ kg}}{1000 \text{ g}} = 0.166 \text{ kg}$$

$F = 0.0450 \text{ N}$

$v_i = 0$

$t = 2.1 \text{ s}$

$v_f = ?$

$$a = \frac{F}{m} = \frac{0.0450 \text{ N}}{0.166 \text{ kg}} = 0.2711 \; \frac{\text{m}}{\text{s}^2}$$

$$a = \frac{v_f - v_i}{t}$$

$$v_f = at + v_i = \left(0.2711 \; \frac{\text{m}}{\text{s}^2} \cdot 2.1 \text{ s}\right) + 0 = 0.57 \; \frac{\text{m}}{\text{s}}$$

12.

$$m = 1.673 \times 10^{-18} \; \mu\text{g} \cdot \frac{1 \text{ g}}{10^6 \; \mu\text{g}} \cdot \frac{1 \text{ kg}}{1000 \text{ g}} = 1.673 \times 10^{-27} \text{ kg}$$

$v_i = 0$

$$v_f = c \cdot 0.0005 = 3.00 \times 10^8 \cdot 0.0005 = 1.50 \times 10^5 \; \frac{\text{m}}{\text{s}}$$

$$t = 455 \text{ ns} \cdot \frac{1 \text{ s}}{10^9 \text{ ns}} = 4.55 \times 10^{-7} \text{ s}$$

$F = ?$

$$a = \frac{v_f - v_i}{t} = \frac{1.50 \times 10^5 \; \frac{\text{m}}{\text{s}} - 0}{4.55 \times 10^{-7} \text{ s}} = 3.30 \times 10^{11} \; \frac{\text{m}}{\text{s}^2}$$

$$a = \frac{F}{m}$$

$$F = ma = 1.673 \times 10^{-27} \text{ kg} \cdot 3.30 \times 10^{11} \; \frac{\text{m}}{\text{s}^2} = 5.52 \times 10^{-16} \text{ N} \cdot \frac{1 \text{ GN}}{10^9 \text{ N}} = 5.52 \times 10^{-25} \text{ GN}$$

13.

$$m = 6.548 \text{ Gg} \cdot \frac{10^9 \text{ g}}{1 \text{ Gg}} \cdot \frac{1 \text{ kg}}{1000 \text{ g}} = 6.548 \times 10^6 \text{ kg}$$

$$v_i = 8.35 \frac{\text{mi}}{\text{hr}} \cdot \frac{5{,}280 \text{ ft}}{1 \text{ mi}} \cdot \frac{0.3048 \text{ m}}{1 \text{ ft}} \cdot \frac{1 \text{ hr}}{60 \text{ min}} \cdot \frac{1 \text{ min}}{60 \text{ s}} = 3.732 \frac{\text{m}}{\text{s}}$$

$$v_f = 0$$

$$t = 0.288 \text{ min} \cdot \frac{60 \text{ s}}{1 \text{ min}} = 17.28 \text{ s}$$

$$F = ?$$

$$a = \frac{v_f - v_i}{t} = \frac{0 - 3.732 \frac{\text{m}}{\text{s}}}{17.28 \text{ s}} = -0.216 \frac{\text{m}}{\text{s}^2}$$

$$a = \frac{F}{m}$$

$$F = ma$$

$$F = 6.548 \times 10^6 \text{ kg} \cdot -0.216 \frac{\text{m}}{\text{s}^2} = 1.41 \times 10^6 \text{ N}$$

14.

$$v_i = 3.5 \frac{\text{cm}}{\text{s}} \cdot \frac{1 \text{ m}}{100 \text{ cm}} = 0.035 \frac{\text{m}}{\text{s}}$$

$$v_f = 18.5 \frac{\text{cm}}{\text{s}} \cdot \frac{1 \text{ m}}{100 \text{ cm}} = 0.185 \frac{\text{m}}{\text{s}}$$

$$t = 220 \text{ ms} \cdot \frac{1 \text{ s}}{1000 \text{ ms}} = 0.22 \text{ s}$$

$$a = ?$$

$$a = \frac{v_f - v_i}{t} = \frac{0.185 \frac{\text{m}}{\text{s}} - 0.035 \frac{\text{m}}{\text{s}}}{0.22 \text{ s}} = 0.68 \frac{\text{m}}{\text{s}^2}$$

15.

$m = 45,500 \text{ kg}$

$v_i = 0 \ \dfrac{\text{m}}{\text{s}}$

$v_f = 55 \ \dfrac{\text{m}}{\text{s}}$

$t = 6.4 \text{ s}$

$a = ?$

$F = ?$

$$a = \frac{v_f - v_i}{t} = \frac{55 \ \dfrac{\text{m}}{\text{s}} - 0}{6.4 \text{ s}} = 8.59 \ \frac{\text{m}}{\text{s}^2}$$

$$\boxed{a = 8.6 \ \frac{\text{m}}{\text{s}^2}}$$

$$a = \frac{F}{m}$$

$$F = ma = 45,500 \text{ kg} \cdot 8.59 \ \frac{\text{m}}{\text{s}^2} = 3.9 \times 10^5 \text{ N}$$

16.

$$m = 8.5 \text{ g} \cdot \frac{1 \text{ kg}}{1000 \text{ g}} = 0.0085 \text{ kg}$$

$$a = 18,500 \ \frac{\text{m}}{\text{s}^2}$$

$F = ?$

$$a = \frac{F}{m}$$

$$F = ma = 0.0085 \text{ kg} \cdot 18,500 \ \frac{\text{m}}{\text{s}^2} = 160 \text{ N}$$

Chapter 4

Note: The values chosen for each activity are examples only, and so answers will vary between students' papers. Students can choose any values for their calculations, as long as their chosen values are within the parameters of each activity.

Activity 1

How does the area of a triangle vary with its height, if all else is held constant?

1. What is the relation for the area of a triangle?

$$A = \frac{hb}{2}$$

2. What are the two key variables that need to be compared in this activity?

A and h

3. Which variable is the independent variable, and which is the dependent variable?

A is dependent; h is independent.

4. After combining and normalizing all non-essential variables and constants, what is the expression relating area to height?

$A \propto h$

5. Select any value other than 2 to use for the base of a triangle. Using this value for the base, choose several values for the height of the triangle and calculate the area of the triangle for each height. Enter all of these in a table of values.

$$A = \frac{hb}{2}$$

$b = 4.0$ m

$h = 2.0$ m, 4.0 m, 6.0 m, 8.0 m, 10.0 m

$$A = \frac{2.0 \text{ m} \cdot 4.0 \text{ m}}{2} = 4.0 \text{ m}^2$$

$$A = \frac{4.0 \text{ m} \cdot 4.0 \text{ m}}{2} = 8.0 \text{ m}^2$$

$$A = \frac{6.0 \text{ m} \cdot 4.0 \text{ m}}{2} = 12 \text{ m}^2$$

$$A = \frac{8.0 \text{ m} \cdot 4.0 \text{ m}}{2} = 16 \text{ m}^2$$

$$A = \frac{10.0 \text{ m} \cdot 4.0 \text{ m}}{2} = 2.0 \times 10^1 \text{ m}^2$$

Height (m)	Area (m²)
2.0	4.0
4.0	8.0
6.0	12
8.0	16
10.0	20.0

Table 1.1. Area of a triangle at selected heights.

6. Prepare a graph of area vs. height using the values you computed in the previous step.

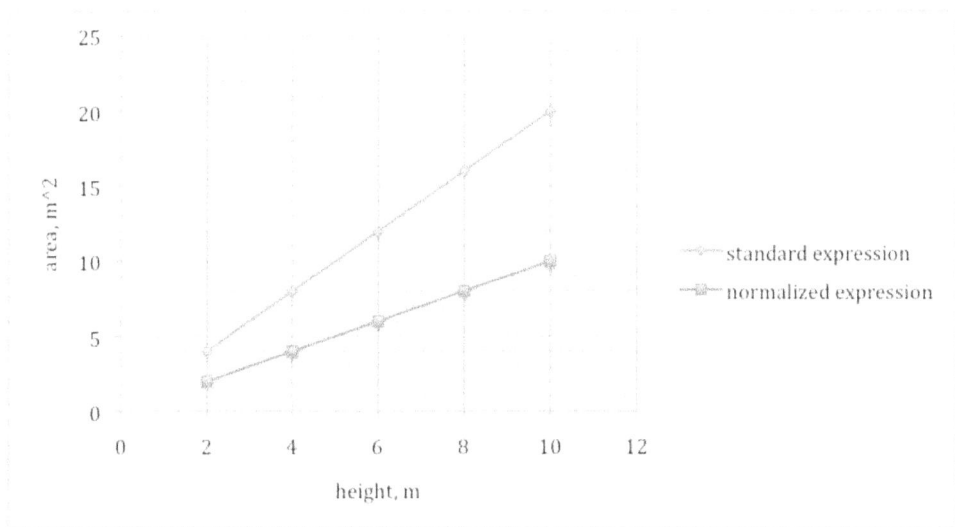

Graph 1.1. Data plot of area vs. height.

7. Compute another table of values for the normalized expression from step 4. Treat the proportional sign as an equals sign for this.

$A \propto h$

$b = 1.0$ m

$h = 2.0$ m, 4.0 m, 6.0 m, 8.0 m, 10.0 m

$A = 2.0$ m^2

$A = 4.0$ m^2

$A = 6.0$ m^2

$A = 8.0$ m^2

$A = 10.0$ m^2

Height (m)	Area (m^2)
2.0	2.0
4.0	6.0
6.0	8.0
8.0	16
10.0	10.0

Table 1.2. Area of a triangle at selected heights using normalized expression.

8. Graph the normalized equation on the same set of coordinate axes you used for the graph in step 6.

See Graph 1.1.

9. Describe the similarities and differences between the two "curves" on your graph.

Both "curves" are actually straight lines. The line for the area of a triangle using the standard expression is steeper than for the normalized expression, but the opposite would be the case if *b* was less than 1.

10. Answer the main question for this activity.

A varies directly with *h*.

Activity 2

How does the area of a circle vary with its radius, if all else is held constant?

1. What is the relation for area of a circle?

$$A = \pi r^2$$

2. What are the two key variables that need to be compared in this activity?

A and r

3. Which variable is the independent variable, and which is the dependent variable?

A is dependent; r is independent.

4. After combining and normalizing all non-essential variables and constants, what is the expression relating area to radius?

$$A \propto r^2$$

5. Choose several values for the radius of the circle and calculate the area of the circle for each radius. Enter all of these in a table of values.

$$A = \pi r^2$$

$r = 2.00$ m, 4.00 m, 6.00 m, 8.00 m, 10.00 m

$$A = \pi(2.00 \text{ m})^2 = 3.14 \cdot 4.00 \text{ m}^2 = 12.6 \text{ m}^2$$

$$A = \pi(4.00 \text{ m})^2 = 3.14 \cdot 16.0 \text{ m}^2 = 50.2 \text{ m}^2$$

$$A = \pi(6.00 \text{ m})^2 = 3.14 \cdot 36.0 \text{ m}^2 = 113 \text{ m}^2$$

$$A = \pi(8.00 \text{ m})^2 = 3.14 \cdot 64.0 \text{ m}^2 = 201 \text{ m}^2$$

$$A = \pi(10.0 \text{ m})^2 = 3.14 \cdot 1.00 \times 10^2 \text{ m}^2 = 314 \text{ m}^2$$

Radius (m)	Area (m²)
2.00	12.6
4.00	50.2
6.00	113
8.00	201
10.00	314

Table 2.1. Area of a circle at selected radii using standard expression.

6. Prepare a graph of area vs. radius using the values you computed in the previous step.

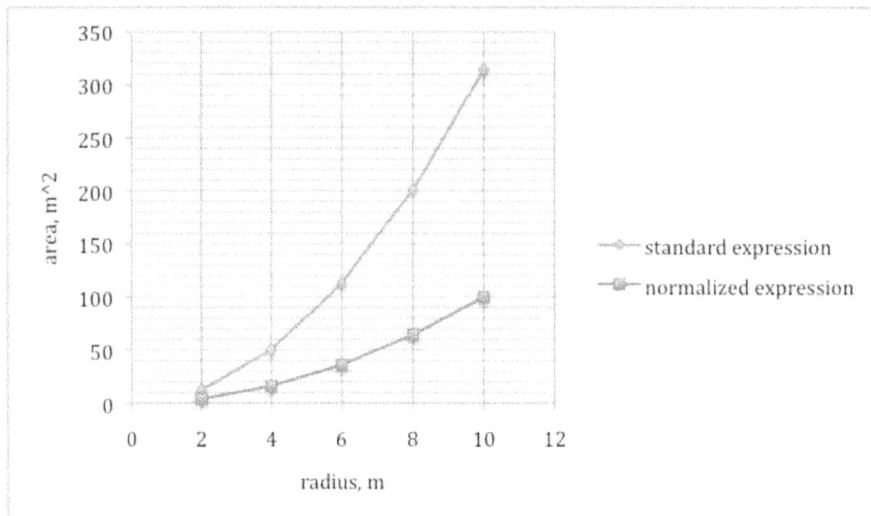

Graph 2.1. Data plot of area vs. radius.

7. Compute another table of values for the normalized expression from step 4. Treat the proportional sign as an equals sign for this.

Radius (m)	Area (m²)
2.00	4.00
4.00	16.0
6.00	36.0
8.00	64.0
10.00	100.0

Table 2.2. Area of a circle at selected radii using normalized expression.

8. Graph the normalized equation on the same set of coordinate axes you used for the graph in step 6.

See Graph 2.1.

9. Describe the similarities and differences between the two curves on your graph.

Both have the same form, but the curve for the normalized expression is stretched horizontally.

10. Answer the main question for this activity.

A varies as the square of r.

Activity 3

How does the volume of a sphere vary with its radius, if all else is held constant?

1. What is the relation for volume of a sphere?

$$V = \frac{4}{3}\pi r^3$$

2. What are the two key variables that need to be compared in this activity?

V and r

3. Which variable is the independent variable, and which is the dependent variable?

V is dependent; r is independent.

4. After combining and normalizing all non-essential variables and constants, what is the expression relating volume to radius?

$$V \propto r^3$$

5. Choose several values for the radius of the sphere and calculate the volume of the sphere for each radius. Enter all of these in a table of values.

$$V = \frac{4}{3}\pi r^3$$

$r = 2.00$ m, 4.00 m , 6.00 m, 8.00 m, 10.00 m

$$V = \frac{4}{3}\pi(2.00 \text{ m})^3 = \frac{4}{3}\cdot 3.14 \cdot 8.00 \text{ m}^3 = 33.5 \text{ m}^3$$

$$V = \frac{4}{3}\pi(4.00 \text{ m})^3 = \frac{4}{3}\cdot 3.14 \cdot 64.0 \text{ m}^3 = 268 \text{ m}^3$$

$$V = \frac{4}{3}\pi(6.00 \text{ m})^3 = \frac{4}{3}\cdot 3.14 \cdot 216 \text{ m}^3 = 904 \text{ m}^3$$

$$V = \frac{4}{3}\pi(8.00 \text{ m})^3 = \frac{4}{3}\cdot 3.14 \cdot 512 \text{ m}^3 = 2{,}140 \text{ m}^3$$

$$V = \frac{4}{3}\pi(10.0 \text{ m})^3 = \frac{4}{3}\cdot 3.14 \cdot 1000 \text{ m}^3 = 4{,}190 \text{ m}^3$$

Radius (m)	Volume (m³)
2.00	33.5
4.00	268
6.00	904
8.00	2,140
10.0	4,190

Table 3.1. Volume of a sphere at selected radii.

6. Prepare a graph of volume vs. radius using the values you computed in the previous step.

Graph 3.1. Data plot of volume versus radius.

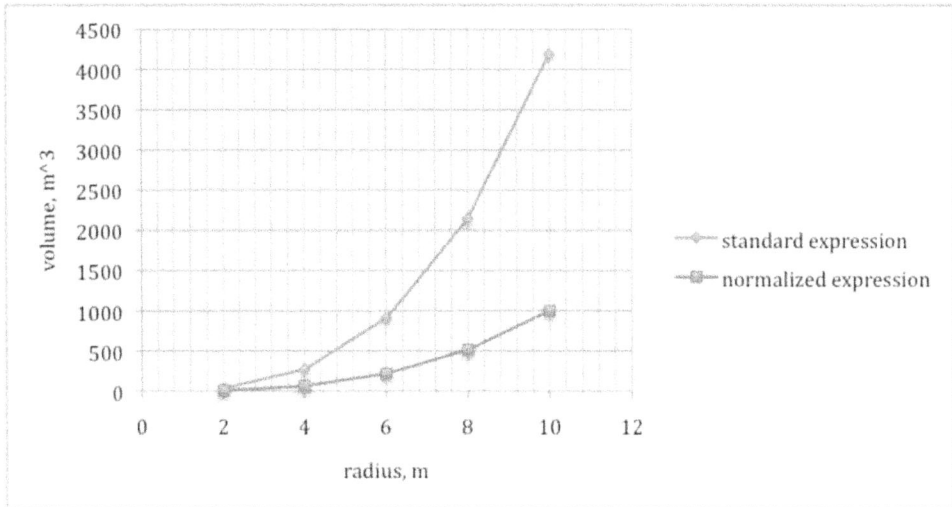

7. Compute another table of values for the normalized expression from step 4. Treat the proportional sign as an equals sign for this.

$$V \propto r^3$$

$r = 2.00$ m, 4.00 m, 6.00 m, 8.00 m, 10.00 m

$$V = (2.00 \text{ m})^3 = 8.00 \text{ m}^3$$
$$V = (4.00)^3 = 64.0 \text{ m}^3$$
$$V = (6.00)^3 = 216 \text{ m}^3$$
$$V = (8.00)^3 = 512 \text{ m}^3$$
$$V = (10.0)^3 = 1.000 \times 10^3 \text{ m}^3$$

Radius (m)	Volume (m³)
2.00	8.00
4.00	64.0
6.00	216
8.00	512
10.0	1000

Table 3.2. Volume of a sphere at selected radii using normalized expression.

8. Graph the normalized equation on the same set of coordinate axes you used for the graph in

step 6.

See Graph 3.1.

9. Describe the similarities and differences between the two "curves" on your graph.

Both curves follow the pattern of $y = kx^3$. The normalized expression shares the same form as the standard expression, but it is stretched horizontally.

10. Answer the main question for this activity.

V varies as the cube of r.

Activity 4

How does the gravitational potential energy of an object vary with its height, if all else is held constant?

1. What are the two key variables that need to be compared in this activity?

E_G and h

2. Which variable is the independent variable, and which is the dependent variable?

E_G is dependent; h is independent.

3. After combining and normalizing all non-essential variables and constants, what is the expression relating to height?

$E_G \propto h$

4. Select a value to use for the mass of the object for this activity. Using this value, choose several values for the height of the object and calculate the of the object for each height. Enter all of these in a table of values.

$E_G = mgh$

$m = 2.00 \text{ kg}$

$g = 9.80 \ \dfrac{\text{m}}{\text{s}^2}$

$h = 2.00 \text{ m}, 4.00 \text{ m}, 6.00 \text{ m}, 8.00 \text{ m}, 10.0 \text{ m}$

$E_G = 2.00 \text{ kg} \cdot 9.80 \ \dfrac{\text{m}}{\text{s}^2} \cdot 2.00 \text{ m} = 39.2 \text{ J}$

$E_G = 2.00 \text{ kg} \cdot 9.80 \ \dfrac{\text{m}}{\text{s}^2} \cdot 4.00 \text{ m} = 78.4 \text{ J}$

$E_G = 2.00 \text{ kg} \cdot 9.80 \ \dfrac{\text{m}}{\text{s}^2} \cdot 6.00 \text{ m} = 118 \text{ J}$

$E_G = 2.00 \text{ kg} \cdot 9.80 \ \dfrac{\text{m}}{\text{s}^2} \cdot 8.00 \text{ m} = 157 \text{ J}$

$E_G = 2.00 \text{ kg} \cdot 9.80 \ \dfrac{\text{m}}{\text{s}^2} \cdot 10.0 \text{ m} = 196 \text{ J}$

Height (m)	Gravitational Potential Energy (J)
2.00	39.2
4.00	78.4
6.00	118
8.00	157
10.0	196

Table 4.1. E_G of an object with mass 2.00 kg at selected heights.

5. Prepare a graph of E_G vs. height using the values you computed in the previous step.

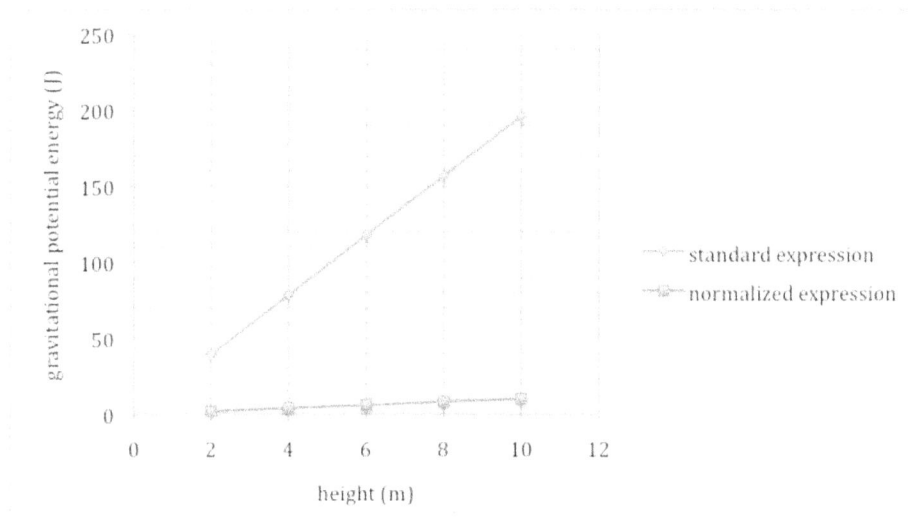

Graph 4.1. Data plot of E_G versus height.

6. Compute another table of values for the normalized expression from step 4. Treat the proportional sign as an equals sign for this.

$$E_G \propto h$$

$$m = 1.00 \text{ kg}$$

$$g = 1.00 \ \frac{m}{s^2}$$

$h = 2.00$ m, 4.00 m, 6.00 m, 8.00 m, 10.0 m

$$E_G = 1.00 \text{ kg} \cdot 1.00 \ \frac{m}{s^2} \cdot 2.00 \text{ m} = 2.00 \text{ J}$$

$$E_G = 1.00 \text{ kg} \cdot 1.00 \ \frac{m}{s^2} \cdot 4.00 \text{ m} = 4.00 \text{ J}$$

$$E_G = 1.00 \text{ kg} \cdot 1.00 \ \frac{m}{s^2} \cdot 6.00 \text{ m} = 6.00 \text{ J}$$

$$E_G = 1.00 \text{ kg} \cdot 1.00 \ \frac{m}{s^2} \cdot 8.00 \text{ m} = 8.00 \text{ J}$$

$$E_G = 1.00 \text{ kg} \cdot 1.00 \ \frac{m}{s^2} \cdot 10.0 \text{ m} = 10.0 \text{ J}$$

Height (m)	Gravitational Potential Energy (J)
2.00	2.00
4.00	4.00
6.00	6.00
8.00	8.00
10.0	10.0

Table 4.22. E_G of an object at selected heights using normalized expression.

7. Graph the normalized equation on the same set of coordinate axes you used for the graph in step 5.

See Graph 4.1.

8. Describe the similarities and differences between the two "curves" on your graph.

Both "curves" are actually straight lines. The normalized expression is stretched horizontally, but both lines follow the pattern of $y = kx$.

9. Answer the main question for this activity.

E_G varies directly with h.

Activity 5

How does the kinetic energy of an object vary with its velocity, if all else is held constant?

1. What are the two key variables that need to be compared in this activity?

E_K and v.

2. Which variable is the independent variable, and which is the dependent variable?

E_K is dependent; v is independent.

3. After combining and normalizing all non-essential variables and constants, what is the expression relating to velocity?

$$E_K \propto v^2$$

4. Select a value other than 2 kg to use for the mass of the object for this activity. Using this value, choose several values for the velocity of the object and calculate the of the object for each velocity. Enter all of these in a table of values.

$$E_K = \frac{1}{2}mv^2$$

$$m = 3.0 \text{ kg}$$

$$v = 2.0 \; \frac{m}{s}, \; 4.0 \; \frac{m}{s}, \; 6.0 \; \frac{m}{s}, \; 8.0 \; \frac{m}{s}, \; 10.0 \; \frac{m}{s}$$

$$E_K = \frac{1}{2} \cdot 3.0 \text{ kg} \cdot (2.0 \; \frac{m}{s})^2 = 6.0 \text{ J}$$

$$E_K = \frac{1}{2} \cdot 3.0 \text{ kg} \cdot (4.0 \; \frac{m}{s})^2 = 24 \text{ J}$$

$$E_K = \frac{1}{2} \cdot 3.0 \text{ kg} \cdot (6.0 \; \frac{m}{s})^2 = 54 \text{ J}$$

$$E_K = \frac{1}{2} \cdot 3.0 \text{ kg} \cdot (8.0 \; \frac{m}{s})^2 = 96 \text{ J}$$

$$E_K = \frac{1}{2} \cdot 3.0 \text{ kg} \cdot (10.0 \; \frac{m}{s})^2 = 1.50 \times 10^2 \text{ J}$$

Velocity ( )	Kinetic Energy (J)
2.0	6.0
4.0	24
6.0	54
8.0	96
10.0	150

Table 5.1. E_K of an object with mass 3.0 kg at selected velocities.

5. Prepare a graph of E_K vs. velocity using the values you computed in the previous step.

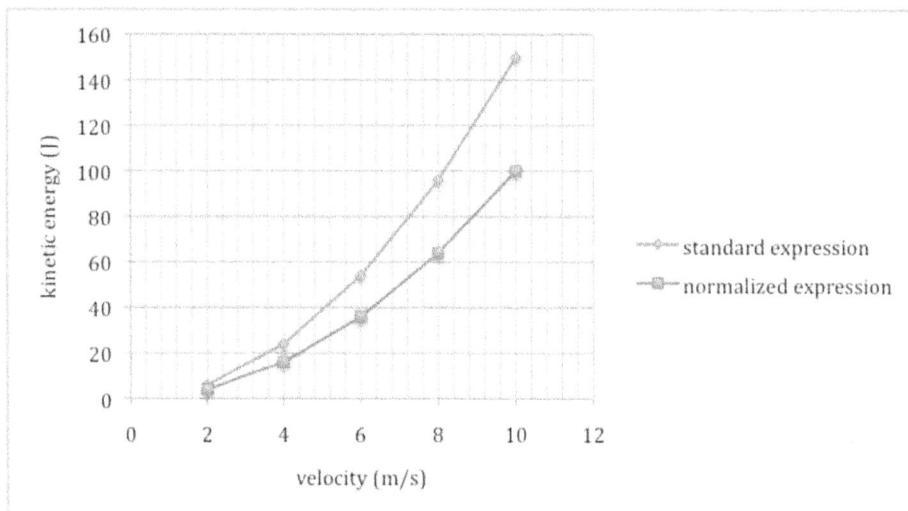

Graph 5.1. Data plot of E_K versus velocity.

6. Compute another table of values for the normalized expression from step 3. Treat the proportional sign as an equals sign for this.

$$E_K \propto v^2$$

$$m = 1.0 \text{ kg}$$

$$v = 2.0 \ \frac{m}{s}, \ 4.0 \ \frac{m}{s}, \ 6.0 \ \frac{m}{s}, \ 8.0 \ \frac{m}{s}, \ 10.0 \ \frac{m}{s}$$

$$E_K = 1.0 \text{ kg} \cdot (2.0 \ \frac{m}{s})^2 = 4.0 \text{ J}$$

$$E_K = 1.0 \text{ kg} \cdot (4.0 \ \frac{m}{s})^2 = 16 \text{ J}$$

$$E_K = 1.0 \text{ kg} \cdot (6.0 \ \frac{m}{s})^2 = 36 \text{ J}$$

$$E_K = 1.0 \text{ kg} \cdot (8.0 \ \frac{m}{s})^2 = 64 \text{ J}$$

$$E_K = 1.0 \text{ kg} \cdot (10.0 \ \frac{m}{s})^2 = 1.00 \times 10^2 \text{ J}$$

Velocity (m/s)	Kinetic Energy (J)
2.0	4.0
4.0	16
6.0	36
8.0	64
10.0	100

Table 5.2. E_K of an object with mass 1.0 kg at selected velocities using normalized expression.

7. Graph the normalized equation on the same set of coordinate axes you used for the graph in step 5.

See Graph 5.

8. Describe the similarities and differences between the two curves on your graph.

Both curves share the same form, but the normalized expression is stretched horizontally. The standard expression curve would vary greatly depending on the value chosen for mass.

9. Answer the main question for this activity.

E_K varies as the square of v.

Activity 6

How does pressure under water vary with depth, if all else is held constant?

1. What are the two key variables that need to be compared in this activity?

P and h

2. Which variable is the independent variable, and which is the dependent variable?

P is dependent; h is independent.

3. After combining and normalizing all non-essential variables and constants, what is the expression relating pressure to depth?

$P \propto h$

4. Using the value for the density of water given above, choose several values for the depth under water and calculate the pressure for each depth. Enter all of these in a table of values.

$P = \rho g h$

$$\rho = 998 \ \frac{kg}{m^3}$$

$$g = 9.80 \ \frac{m}{s^2}$$

$h = 0.020$ m, 0.040 m, 0.060 m, 0.080 m, 0.100 m

$$P = 998 \ \frac{kg}{m^3} \cdot 9.80 \ \frac{m}{s^2} \cdot 0.020 \ m = 196 \ Pa$$

$$P = 998 \ \frac{kg}{m^3} \cdot 9.80 \ \frac{m}{s^2} \cdot 0.040 \ m = 391 \ Pa$$

$$P = 998 \ \frac{kg}{m^3} \cdot 9.80 \ \frac{m}{s^2} \cdot 0.060 \ m = 587 \ Pa$$

$$P = 998 \ \frac{kg}{m^3} \cdot 9.80 \ \frac{m}{s^2} \cdot 0.080 \ m = 782 \ Pa$$

$$P = 998 \ \frac{kg}{m^3} \cdot 9.80 \ \frac{m}{s^2} \cdot 0.100 \ m = 978 \ Pa$$

Depth (m)	Pressure (Pa)
0.020	196
0.040	391
0.060	587
0.080	782
0.100	978

Table 6.1. Pressure under water at selected depths.

5. Prepare a graph of pressure vs. depth using the values you computed in the previous step.

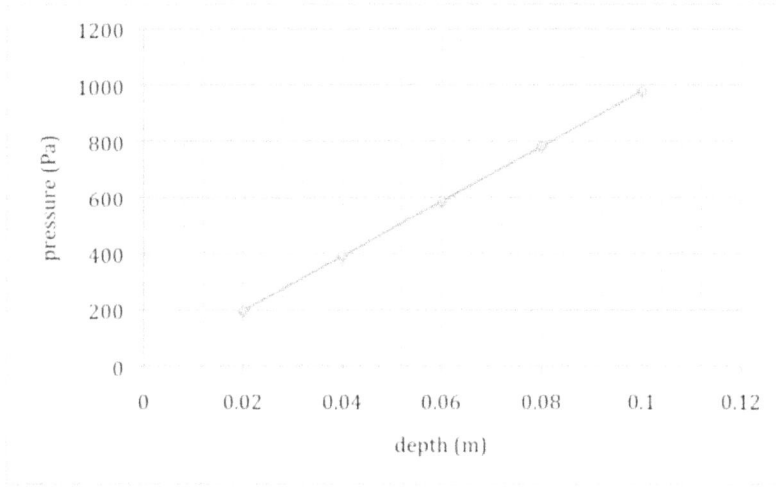

Graph 6.1. Data plot of pressure under water versus depth using standard expression.

6. Compute another table of values for the normalized expression from step 3. Treat the proportional sign as an equals sign for this.

$P \propto h$

$$\rho = 1 \ \frac{kg}{m^3}$$

$$g = 1.00 \ \frac{m}{s^2}$$

$h = 0.020$ m, 0.040 m, 0.060 m, 0.080 m, 0.100 m

$$P = 1 \ \frac{kg}{m^3} \cdot 1 \ \frac{m}{s^2} \cdot 0.020 \ m = 0.020 \ Pa$$

$$P = 1 \ \frac{kg}{m^3} \cdot 1 \ \frac{m}{s^2} \cdot 0.040 \ m = 0.040 \ Pa$$

$$P = 1 \ \frac{kg}{m^3} \cdot 1 \ \frac{m}{s^2} \cdot 0.060 \ m = 0.060 \ Pa$$

$$P = 1 \ \frac{kg}{m^3} \cdot 1 \ \frac{m}{s^2} \cdot 0.080 \ m = 0.080 \ Pa$$

$$P = 1 \ \frac{kg}{m^3} \cdot 1 \ \frac{m}{s^2} \cdot 0.100 \ m = 0.100 \ Pa$$

Depth (m)	Pressure (Pa)
0.020	0.020
0.040	0.040
0.060	0.060
0.080	0.080
0.100	0.100

Table 6.2. Pressure under water at selected depths using normalized expression.

7. Graph the normalized equation on a separate set of coordinate axes you used for the graph in step 5.

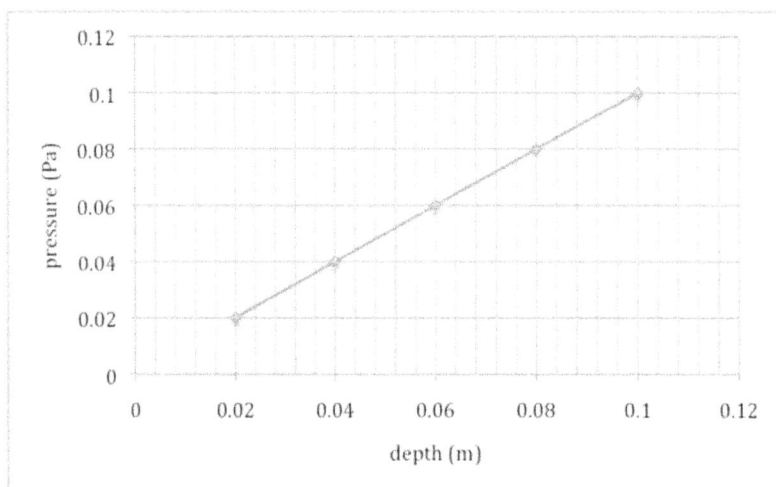

Graph 6.2. Data plot of pressure versus depth using normalized expression.

8. Describe the similarities and differences between the two "curves" on your graph.

Both are lines, but the standard expression is much, much steeper due to the high density of water. It would be virtually meaningless to graph them on the same set of axes, since the normalized expression would almost look like it was a horizontal line.

9. Answer the main question for this activity.

Pressure under water varies directly with the depth.

Activity 7

How does the force of gravitational attraction vary with the distance between the centers of two objects?

1. What are the two key variables that need to be compared in this activity?

F and d

2. Which variable is the independent variable, and which is the dependent variable?

F is dependent; d is independent.

3. After combining and normalizing all non-essential variables and constants, what is the expression relating the gravitational force between two objects to the distance between them?

$$F \propto \frac{1}{d^2}$$

4. Select values to use for the two masses of the objects for this activity. Using these values, choose several values for the distance between the objects and calculate the force of attraction for each distance. Enter all of these in a table of values.

$$F = G\frac{m_1 m_2}{d^2}$$

$$G = 6.67 \times 10^{-11} \ \frac{\text{Nm}^2}{\text{kg}^2}$$

$m_1 = 2.5$ kg

$m_2 = 3.0$ kg

$d = 2.0$ m, 4.0 m, 6.0 m, 8.0 m, 10.0 m

$$F = 6.67 \times 10^{-11} \ \frac{\text{Nm}^2}{\text{kg}^2} \cdot \frac{2.5 \ \text{kg} \cdot 3.0 \ \text{kg}}{(2.0 \ \text{m})^2} = 130 \times 10^{-12} \ \text{N}$$

$$F = 6.67 \times 10^{-11} \ \frac{\text{Nm}^2}{\text{kg}^2} \cdot \frac{2.5 \ \text{kg} \cdot 3.0 \ \text{kg}}{(4.0 \ \text{m})^2} = 31 \times 10^{-12} \ \text{N}$$

$$F = 6.67 \times 10^{-11} \ \frac{\text{Nm}^2}{\text{kg}^2} \cdot \frac{2.5 \ \text{kg} \cdot 3.0 \ \text{kg}}{(6.0 \ \text{m})^2} = 14 \times 10^{-12} \ \text{N}$$

$$F = 6.67 \times 10^{-11} \ \frac{\text{Nm}^2}{\text{kg}^2} \cdot \frac{2.5 \ \text{kg} \cdot 3.0 \ \text{kg}}{(8.0 \ \text{m})^2} = 7.8 \times 10^{-12} \ \text{N}$$

$$F = 6.67 \times 10^{-11} \ \frac{\text{Nm}^2}{\text{kg}^2} \cdot \frac{2.5 \ \text{kg} \cdot 3.0 \ \text{kg}}{(10.0 \ \text{m})^2} = 5.0 \times 10^{-12} \ \text{N}$$

Distance (m)	Gravitational force (N)
2.0	130×10^{-12}
4.0	31×10^{-12}
6.0	14×10^{-12}
8.0	7.8×10^{-12}
10.0	5.0×10^{-12}

Table 7.1. Gravitational force between two objects of mass 2.5 and 3.0 kg at selected distances.

5. Prepare a graph of force vs. distance using the values you computed in the previous step.

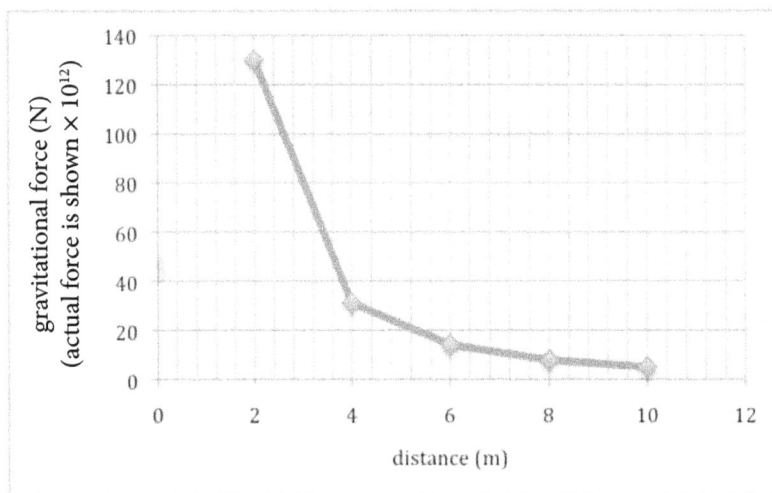

Graph 7.1. Data plot of gravitational force versus distance between two objects of mass 2.5 and 3.0 kg.

6. Compute another table of values for the normalized expression from step 3. Treat the proportional sign as an equals sign for this.

$$F \propto \frac{1}{d^2}$$

$$G = 1 \frac{\text{Nm}^2}{\text{kg}^2}$$

$$m_1 = 1.0 \text{ kg}$$

$$m_2 = 1.0 \text{ kg}$$

$$d = 2.0 \text{ m, } 4.0 \text{ m, } 6.0 \text{ m, } 8.0 \text{ m, } 10.0 \text{ m}$$

$$F = 1 \frac{\text{Nm}^2}{\text{kg}^2} \cdot \frac{1.0 \text{ kg} \cdot 1.0 \text{ kg}}{(2.0 \text{ m})^2} = 0.25 \text{ N}$$

$$F = 1 \frac{\text{Nm}^2}{\text{kg}^2} \cdot \frac{1.0 \text{ kg} \cdot 1.0 \text{ kg}}{(4.0 \text{ m})^2} = 0.06 \text{ N}$$

$$F = 1 \frac{\text{Nm}^2}{\text{kg}^2} \cdot \frac{1.0 \text{ kg} \cdot 1.0 \text{ kg}}{(6.0 \text{ m})^2} = 0.03 \text{ N}$$

$$F = 1 \frac{\text{Nm}^2}{\text{kg}^2} \cdot \frac{1.0 \text{ kg} \cdot 1.0 \text{ kg}}{(8.0 \text{ m})^2} = 0.02 \text{ N}$$

$$F = 1 \frac{\text{Nm}^2}{\text{kg}^2} \cdot \frac{1.0 \text{ kg} \cdot 1.0 \text{ kg}}{(10.0 \text{ m})^2} = 0.01 \text{ N}$$

Distance (m)	Gravitational force (N)
2.0	0.25
4.0	0.06
6.0	0.03
8.0	0.02
10.0	0.01

Table 7.2. Gravitational force between two objects at selected distances using normalized expression.

7. Graph the normalized equation on a separate set of coordinate axes you used for the graph in step 5.

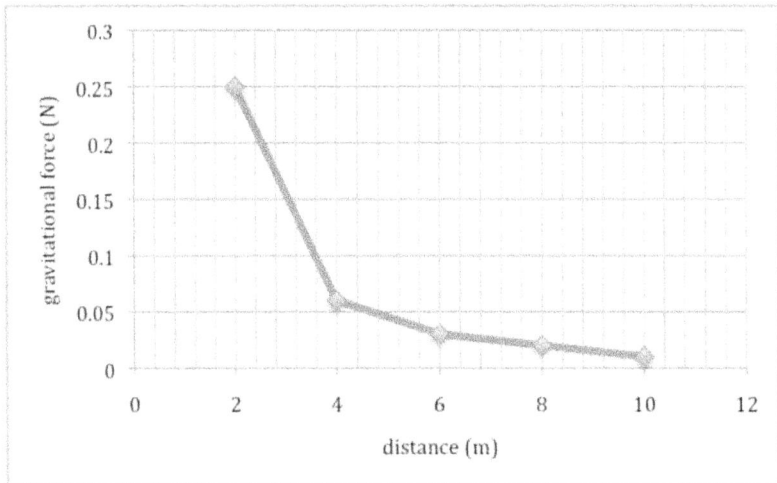

Graph 7.2. Data plot of gravitational force versus distance between two objects using normalized expression.

8. Describe the similarities and differences between the two curves on your graph.

The curves are almost identical; however, the scales are vastly different.

9. Answer the main question for this activity.

The force of gravitational attraction varies as the inverse square of distance between two objects.

Activity 8

Note: The values in the first, third, and fifth columns presented below were taken from the "Demo Data" found in *Favorite Experiments*, Part 2, Demos 3 on Variation and Proportion. Details for the book can be found on novarescienceandmath.com.

First, enter the data from the demo in your table.

T_C (C)	T_K (K)	V_{buret} (mL = cm^3)	ΔV (cm^3)	$V_{experimental}$	$V_{predicted}$
3.7	$T_i =$	24.7	O	$V_i =$	$V_i =$
9.8		24.6			
15.0		24.5			
18.4		24.4			
23.5		24.3			
28.8		24.2			
33.8		24.1			
37.0		24.0			

Table 8.1. Initial table for Charles' Law activity with demo data.

Next, convert all of the temperature values from degrees Celsius to kelvins.

$$T_K = T_C + 273.2$$
$$T_K = 3.7°C + 273.2 = 276.9 \text{ K}$$
$$T_K = 9.8°C + 273.2 = 283.0 \text{ K}$$
$$T_K = 15.0°C + 273.2 = 288.2 \text{ K}$$
$$T_K = 18.4°C + 273.2 = 291.6 \text{ K}$$
$$T_K = 23.5°C + 273.2 = 296.7 \text{ K}$$
$$T_K = 28.8°C + 273.2 = 302.0 \text{ K}$$
$$T_K = 33.8°C + 273.2 = 307.0 \text{ K}$$
$$T_K = 37.0°C + 273.2 = 310.2 \text{ K}$$

Enter all of the temperatures in kelvins in the table. Next, we will obtain the values for ΔV.

$\Delta V = (\text{first reading of } V_{buret}) - V_{buret}$

$\Delta V = 24.7 \text{ cm}^3 - 24.7 \text{ cm}^3 = 0 \text{ cm}^3$

$\Delta V = 24.7 \text{ cm}^3 - 24.6 \text{ cm}^3 = 0.1 \text{ cm}^3$

$\Delta V = 24.7 \text{ cm}^3 - 24.5 \text{ cm}^3 = 0.2 \text{ cm}^3$

$\Delta V = 24.7 \text{ cm}^3 - 24.4 \text{ cm}^3 = 0.3 \text{ cm}^3$

$\Delta V = 24.7 \text{ cm}^3 - 24.3 \text{ cm}^3 = 0.4 \text{ cm}^3$

$\Delta V = 24.7 \text{ cm}^3 - 24.2 \text{ cm}^3 = 0.5 \text{ cm}^3$

$\Delta V = 24.7 \text{ cm}^3 - 24.1 \text{ cm}^3 = 0.6 \text{ cm}^3$

$\Delta V = 24.7 \text{ cm}^3 - 24.0 \text{ cm}^3 = 0.7 \text{ cm}^3$

Next we need to determine the value of the initial air volume in the buret, V_i.

$$V_{experimental} = V_i + \Delta V$$

$$V_i = ?$$

$$V = \left(\frac{V_i}{T_i}\right) T$$

$$V_i + \Delta V = \left(\frac{V_i}{T_i}\right) T$$

$$V_i = \frac{\Delta V}{\left(\dfrac{T}{T_i} - 1\right)}$$

$$T_K = T_C + 273.2$$

$$T_K = 35°C + 273.2 = 308.2 \text{ K}$$

$$V_i = \frac{0.7 \text{ cm}^3}{\left(\dfrac{310.2 \text{ K}}{276.9 \text{ K}} - 1\right)} = 5.82 \text{ cm}^3$$

Finally, now that we have V_i, we will use it with T_i and the values of T to calculate the predicted value of the volume for each temperature.

$$V_{predicted} = \left(\frac{V_i}{T_i}\right) T$$

$$V_{predicted} = \left(\frac{5.82\ \text{cm}^3}{276.9\ \text{K}}\right) \cdot 283.0\ \text{K} = 5.95\ \text{cm}^3$$

$$V_{predicted} = \left(\frac{5.82\ \text{cm}^3}{276.9\ \text{K}}\right) \cdot 288.2\ \text{K} = 6.06\ \text{cm}^3$$

$$V_{predicted} = \left(\frac{5.82\ \text{cm}^3}{276.9\ \text{K}}\right) \cdot 291.6\ \text{K} = 6.13\ \text{cm}^3$$

$$V_{predicted} = \left(\frac{5.82\ \text{cm}^3}{276.9\ \text{K}}\right) \cdot 296.7\ \text{K} = 6.24\ \text{cm}^3$$

$$V_{predicted} = \left(\frac{5.82\ \text{cm}^3}{276.9\ \text{K}}\right) \cdot 302.0\ \text{K} = 6.35\ \text{cm}^3$$

$$V_{predicted} = \left(\frac{5.82\ \text{cm}^3}{276.9\ \text{K}}\right) \cdot 307.0\ \text{K} = 6.45\ \text{cm}^3$$

$$V_{predicted} = \left(\frac{5.82\ \text{cm}^3}{276.9\ \text{K}}\right) \cdot 310.2\ \text{K} = 6.52\ \text{cm}^3$$

T_C (C)	T_K (K)	V_{buret} (mL = cm^3)	ΔV (cm^3)	$V_{experimental}$	$V_{predicted}$
3.7	T_i=276.9	24.7	O	V_i=5.82	V_i=5.82
9.8	283.0	24.6	.1	5.92	5.95
15.0	288.2	24.5	.2	6.01	6.06
18.4	291.6	24.4	.3	6.12	6.13
23.5	296.7	24.3	.4	6.22	6.24
28.8	302.0	24.2	.5	6.32	6.35
33.8	307.0	24.1	.6	6.42	6.45
37.0	310.2	24.0	.7	6.52	6.52

Table 8.2. Completed table for Charles' Law Activity.

The result of our work will be a graph of volume vs. temperature, showing both predicted and experimental values of volume for each value of temperature.

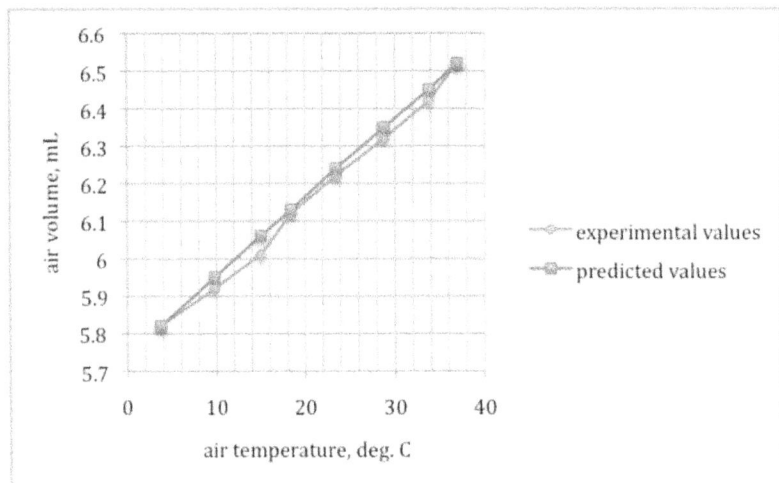

Graph 8.1 Volume vs. temperature for Charles' Law demonstration.

Activity 9

How does the volume of a gas vary with its pressure?

1. What are the two key variables that need to be compared in this activity?

V and *P*

2. Which of these is the independent variable, and which is the dependent variable?

V is dependent; *P* is independent.

3. After combining and normalizing all non-essential variables and constants, what is the expression relating volume to pressure?

$$V \propto \frac{1}{P}$$

4. How are these variables related to one another?

V varies inversely with *P*.

5. Your job is to use Boyle's Law to calculate the volume of a balloon at various pressures from atmospheric pressure at the surface of the water down to the pressure at a depth of 50.0 m. So your lowest pressure value is at the surface; the highest pressure value is the pressure at 50.0 m deep. You can pick several pressures in between and calculate the volume for them as well.

Graph the changes in the volume as the pressure is increased from atmospheric pressure at the surface to pressure at final depth.

$$V = \frac{V_i P_i}{P}$$

$$P_i = 101.3 \text{ kPa} \cdot \frac{1000 \text{ Pa}}{1 \text{ kPa}} = 1.013 \times 10^5 \text{ Pa}$$

$$V_i = 0.0071 \text{ m}^3$$

$$P = 1.013 \times 10^5 \text{ Pa}, \, 2.013 \times 10^5 \text{ Pa}, \, 3.013 \times 10^5 \text{ Pa}, \, 4.013 \times 10^5, \, 5.903 \times 10^5 \text{ Pa}$$

$$V = \frac{0.0071 \text{ m}^3 \cdot 1.013 \times 10^5 \text{ Pa}}{1.013 \times 10^5 \text{ Pa}} = 0.0071 \text{ m}^3$$

$$V = \frac{0.0071 \text{ m}^3 \cdot 1.013 \times 10^5 \text{ Pa}}{2.013 \times 10^5 \text{ Pa}} = 0.0036 \text{ m}^3$$

$$V = \frac{0.0071 \text{ m}^3 \cdot 1.013 \times 10^5 \text{ Pa}}{3.013 \times 10^5 \text{ Pa}} = 0.0024 \text{ m}^3$$

$$V = \frac{0.0071 \text{ m}^3 \cdot 1.013 \times 10^5 \text{ Pa}}{4.013 \times 10^5 \text{ Pa}} = 0.0018 \text{ m}^3$$

$$V = \frac{0.0071 \text{ m}^3 \cdot 1.013 \times 10^5 \text{ Pa}}{5.903 \times 10^5 \text{ Pa}} = 0.0012 \text{ m}^3$$

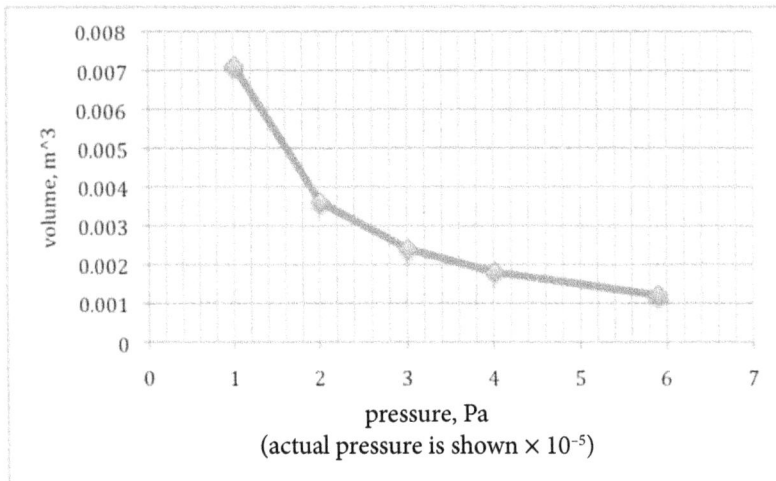

Graph 9.1. Data plot of volume versus pressure.

6. Find out how big this balloon is at this depth. Take the final volume you calculated for a depth of 50.0 m and use the equation below to calculate the radius of the balloon at 50.0 m depth. Then double it to get the diameter. Give your final result in inches.

$$r = \left(\frac{3V}{4\pi} \right)^{\frac{1}{3}}$$

$$r = \left(\frac{3 \cdot 0.00122 \text{ m}}{4\pi} \right)^{\frac{1}{3}} = 0.06629 \text{ m} \cdot \frac{1 \text{ ft}}{0.3048 \text{ m}} \cdot \frac{12 \text{ in}}{1 \text{ ft}} = 2.61 \text{ in}$$

$$D = 2r = 2 \cdot 2.61 \text{ in} = 5.2 \text{ in}$$

Chapter 5

Classroom Examples

1.

$m = 1.00 \times 10^5$ kg

$h = 240 \text{ ft} \cdot \dfrac{0.3048 \text{ m}}{1 \text{ ft}} = 73.15 \text{ m}$

$g = 9.80 \dfrac{\text{m}}{\text{s}^2}$

$E_G = ?$

$E_G = mgh = 1.00 \times 10^5 \text{ kg} \cdot 9.80 \dfrac{\text{m}}{\text{s}^2} \cdot 73.15 \text{ m} = 72{,}000{,}000 \text{ J}$

2.

$m = 25 \text{ g} \cdot \dfrac{1 \text{ kg}}{1000 \text{ g}} = 0.025 \text{ kg}$

$v = 556 \dfrac{\text{ft}}{\text{s}} \cdot \dfrac{0.3048 \text{ m}}{1 \text{ ft}} = 169.5 \dfrac{\text{m}}{\text{s}}$

$E_K = ?$

$E_K = \dfrac{1}{2} mv^2 = \dfrac{1}{2} \cdot 0.025 \text{ kg} \cdot \left(169.5 \dfrac{\text{m}}{\text{s}}\right)^2 = 360 \text{ J}$

3.

$d = 75 \text{ cm} \cdot \dfrac{1 \text{ m}}{100 \text{ cm}} = 0.75 \text{ m}$

$m = 12{,}500 \text{ g} \cdot \dfrac{1 \text{ kg}}{1000 \text{ g}} = 12.5 \text{ kg}$

$W = ?$

$W = Fd$

$F_w = mg = 12.5 \text{ kg} \cdot 9.80 \dfrac{\text{m}}{\text{s}^2} = 122.5 \text{ kg}$

$W = 122.5 \text{ kg} \cdot 0.75 \text{ m} = 92 \text{ J}$

4.

$m = 12.5$ kg

$h = 0.75$ m

$E_G = mgh = 0.75 \text{ m} \cdot 9.80 \dfrac{\text{m}}{\text{s}^2} \cdot 12.5 \text{ kg} = 92 \text{ J}$

5.

$m = 12.5$ kg

$h_i = 0.75$ m

$h_f = 0$

$v_i = 0$

$v_f = ?$

$E_{Gi} + E_{Ki} = E_{Gf} + E_{Kf}$

$E_{Kf} = E_{Gi} + E_{Ki} - E_{Gf} = 92 \text{ J} + 0 - 0 = 92 \text{ J}$

$v_f = \sqrt{\dfrac{2E_{Kf}}{m}} = \sqrt{\dfrac{2 \cdot 92 \text{ J}}{12.5 \text{ kg}}} = 3.8 \dfrac{\text{m}}{\text{s}}$

6.

$$m = 255.8 \text{ g} \cdot \frac{1 \text{ kg}}{1000 \text{ g}} = 0.2558 \text{ kg}$$

$$h_i = 10.4 \text{ ft} \cdot \frac{0.3048 \text{ m}}{1 \text{ ft}} = 3.1699 \text{ m}$$

$$E_{Gi} = ?$$

$$E_{Gi} = mgh_i = 0.2558 \text{ kg} \cdot 9.80 \ \frac{\text{m}}{\text{s}^2} \cdot 3.1699 \text{ m} = 7.95 \text{ J}$$

$$h_f = 0$$

$$v_i = 0$$

$$v_f = ?$$

$$E_{Gi} + E_{Ki} = E_{Gf} + E_{Kf}$$

$$E_{Ki} = 0$$

$$E_{Gf} = 0$$

$$E_{Kf} = E_{Gi} + E_{Ki} - E_{Gf} = 7.95 \text{ J} + 0 \text{ J} - 0 \text{ J} = 7.95 \text{ J}$$

$$v_f = \sqrt{\frac{2E_{Kf}}{m}} = \sqrt{\frac{2 \cdot 7.95 \text{ J}}{0.2558 \text{ kg}}} = 7.88 \ \frac{\text{m}}{\text{s}}$$

Energy Calculations Set 1

1.

$$m = 1.31 \times 10^3 \text{ kg}$$

$$h = 177.44 \text{ ft} \cdot \frac{0.3048 \text{ m}}{1 \text{ ft}} = 54.084 \text{ m}$$

$$E_G = ?$$

$$E_G = mgh = 1.31 \times 10^3 \text{ kg} \cdot 9.80 \ \frac{\text{m}}{\text{s}^2} \cdot 54.084 \text{ m} = 694,000 \text{ J}$$

2.

$$m = 2,345 \text{ kg}$$

$$v = 31 \ \frac{\text{mi}}{\text{hr}} \cdot \frac{5,280 \text{ ft}}{1 \text{ mi}} \cdot \frac{0.3048 \text{ m}}{1 \text{ ft}} \cdot \frac{1 \text{ hr}}{60 \text{ min}} \cdot \frac{1 \text{ min}}{60 \text{ s}} = 13.858 \ \frac{\text{m}}{\text{s}}$$

$$E_K = \frac{1}{2}mv^2 = \frac{1}{2} \cdot 2,345 \text{ kg} \cdot (13.858 \ \frac{\text{m}}{\text{s}})^2 = 230,000 \text{ J}$$

3.

$$d = 61.7 \text{ cm} \cdot \frac{1 \text{ m}}{100 \text{ cm}} = 0.617 \text{ m}$$

$m = 17.5 \text{ kg}$

$W = ?$

$$a = \frac{F}{m}$$

$$F_w = mg = 17.5 \text{ kg} \cdot 9.80 \ \frac{m}{s^2} = 171.5 \text{ N}$$

$$W = F_w d = 171.5 \text{ N} \cdot 0.617 \text{ m} = 106 \text{ J}$$

4.

$$h = 61.7 \text{ cm} \cdot \frac{1 \text{ m}}{100 \text{ cm}} = 0.617 \text{ m}$$

$m = 17.5 \text{ kg}$

$E_G = ?$

$$E_G = mgh = 17.5 \text{ kg} \cdot 9.80 \ \frac{m}{s^2} \cdot 0.617 \text{ m} = 106 \text{ J}$$

5.

$m = 17.5 \text{ kg}$

$h_f = 0.617 \text{ m}$

$h_i = 0$

$v_i = 0$

$v_f = ?$

$$E_{Gi} + E_{Ki} = E_{Gf} + E_{Kf}$$

$$E_{Kf} = E_{Gi} + E_{Ki} - E_{Gf} = 106 \text{ J} + 0 \text{ J} - 0 \text{ J} = 106 \text{ J}$$

$$v_f = \sqrt{\frac{2 E_{Kf}}{m}} = \sqrt{\frac{2 \cdot 106 \text{ J}}{17.5 \text{ kg}}} = 12.1 \ \frac{m}{s} = 3.48 \ \frac{m}{s}$$

6.

$$m = 122 \text{ g} \cdot \frac{1 \text{ kg}}{1000 \text{ g}} = 0.122 \text{ kg}$$

$$v_i = 13.75 \ \frac{m}{s}$$

$$v_f = 0$$

$$h_i = 0$$

$$h_f = ?$$

$$E_{Ki} = \frac{1}{2}mv^2 = \frac{1}{2} \cdot 0.122 \text{ kg} \cdot (13.75 \ \frac{m}{s})^2 = 11.53 \text{ J}$$

$$E_{Gi} + E_{Ki} = E_{Gf} + E_{Kf}$$

$$E_{Gf} = E_{Gi} + E_{Ki} - E_{Kf} = 0 + 11.53 \text{ J} - 0 = 11.53 \text{ J}$$

$$E_{Gf} = mgh_f$$

$$h_f = \frac{E_{Gf}}{mg} = \frac{11.53 \text{ J}}{0.122 \text{ kg} \cdot 9.8 \ \frac{m}{s^2}} = 9.65 \text{ m}$$

7.

$$m = 325 \text{ g} \cdot \frac{1 \text{ kg}}{1000 \text{ g}} = 0.325 \text{ kg}$$

$$h_i = 36.1 \text{ m}$$

$$h_f = 0$$

$$v_i = 0$$

$$v_f = ?$$

$$E_{Gi} = mgh_i = 0.325 \text{ kg} \cdot 9.80 \ \frac{m}{s^2} \cdot 36.1 \text{ m} = 114.98 \text{ J}$$

$$E_{Gi} + E_{Ki} = E_{Gf} + E_{Kf}$$

$$E_{Kf} = E_{Gi} + E_{Ki} - E_{Gf} = 114.98 \text{ J} + 0 - 0$$

$$E_{Kf} = 114.98 \text{ J}$$

$$E_{Kf} = \frac{1}{2}mv^2$$

$$v_f = \sqrt{\frac{2E_{Kf}}{m}} = \sqrt{\frac{2 \cdot 114.98 \text{ J}}{0.325 \text{ kg}}} = 26.6 \ \frac{m}{s}$$

8.

$F = 735 \text{ N}$

$d = 26 \text{ m}$

$W = ?$

$W = Fd = 735 \text{ N} \cdot 26 \text{ m} = 19,000 \text{ J}$

Energy Calculations Set 2

1.a.

$F_w = 20 \cdot 80.0 \text{ lb} \cdot \dfrac{4.45 \text{ N}}{1 \text{ lb}} = 7,120 \text{ N}$

$h = 8.5 \text{ m}$

$g = 9.80 \ \dfrac{\text{m}}{\text{s}^2}$

$m = ?$

$F_w = mg$

$m = \dfrac{F_w}{g}$

$m = \dfrac{7,120 \text{ N}}{9.80 \ \dfrac{\text{m}}{\text{s}^2}} = 727 \text{ kg}$

1.b

$F = 7,120 \text{ N}$

$d = 8.5 \text{ m}$

$W = Fd = 7,120 \text{ N} \cdot 8.5 \text{ m}$

$W = 6.0 \times 10^4 \text{ J}$

1.c.

$m = 727 \text{ kg}$

$g = 9.80 \ \dfrac{\text{m}}{\text{s}^2}$

$h = 8.5 \text{ m}$

$E_G = mgh$

$E_G = 727 \text{ kg} \cdot 9.80 \ \dfrac{\text{m}}{\text{s}^2} \cdot 8.5 \text{ m} = 6.0 \times 10^4 \text{ J}$

1.d.

$m = 767$ kg

$v_i = 0$

$E_{Ki} = ?$

$E_{Ki} = \dfrac{1}{2}mv_i^{\,2} = \dfrac{1}{2} \cdot 767 \text{ kg} \cdot 0^2 = 0 \text{ J}$

$E_{Gi} = 6.0 \times 10^4$ J

$E_{Gf} = 0$

$E_{Kf} = ?$

$E_{Gi} + E_{Ki} = E_{Gf} + E_{Kf}$

$E_{Kf} = E_{Gi} + E_{Ki} - E_{Gf} = 6.0 \times 10^4 \text{ J} + 0 - 0 = 6.0 \times 10^4 \text{ J}$

1.e.

$m = 767$ kg

$E_{Kf} = 6.0 \times 10^4$ J

$v_f = ?$

$E_{Kf} = \dfrac{1}{2}mv_f^{\,2}$

$v_f = \sqrt{\dfrac{2E_{Kf}}{m}} = \sqrt{\dfrac{2 \cdot 6.0 \times 10^4 \text{ J}}{767 \text{ kg}}} = 13 \ \dfrac{\text{m}}{\text{s}}$

2.a.

$$F_W = 3,193 \text{ lb} \cdot \frac{4.45 \text{ N}}{1 \text{ lb}} = 14,209 \text{ N}$$

$m = ?$

$F_W = mg$

$$m = \frac{F_W}{g}$$

$$m = \frac{14,209 \text{ N}}{9.80 \, \dfrac{\text{m}}{\text{s}^2}} = 1,450 \text{ kg}$$

$h = 16 \text{ m}$

$$E_G = mgh = 1,450 \text{ kg} \cdot 9.80 \, \frac{\text{m}}{\text{s}^2} \cdot 16 \text{ m} = 227,360 \text{ J}$$

$E_G = 230,000 \text{ J}$

2.c.

$E_{Kf} = ?$

$E_{G_i} + E_{Ki} = E_{Gf} + E_{Kf}$

$E_{Kf} = E_{G_i} + E_{Ki} - E_{Gf}$

$E_{Kf} = 230,000 \text{ J} + 0 - 0 = 230,000 \text{ J}$

2.d.

$E_{Kf} = 230,000 \text{ J}$

$m = 1,450 \text{ kg}$

$v_f = ?$

$$v_f = \sqrt{\frac{2E_{Kf}}{m}} = \sqrt{\frac{2 \cdot 230,000 \text{ J}}{1,450 \text{ kg}}} = 18 \, \frac{\text{m}}{\text{s}}$$

3.a.

$m = 1,450 \text{ kg}$

$h = 8 \text{ m}$

$E_{G8(at \, h \, = \, 8)} = ?$

$$E_{G8} = mgh = 1,450 \text{ kg} \cdot 9.80 \, \frac{\text{m}}{\text{s}^2} \cdot 8 \text{ m} = 113,680 \text{ J}$$

$E_{G8} = 114,000 \text{ J}$

3.b.

$m = 1,450 \text{ kg}$

$h = 8 \text{ m}$

$E_{G8} = 113,680 \text{ J}$

$E_{K8} = ?$

$E_{G8} + E_{K8} = E_{Gf} + E_{Kf}$

$E_{K8} = E_{Gf} + E_{Kf} - E_{G8} = 0 + 227,360 \text{ J} - 113,680 \text{ J} = 114,000 \text{ J}$

3.c.

$E_{K8} = 114,000 \text{ J}$

$v_8 = ?$

$m = 1,450 \text{ kg}$

$v_8 = \sqrt{\dfrac{2E_{K8}}{m}} = \sqrt{\dfrac{2 \cdot 114,000 \text{ J}}{1,450 \text{ kg}}} = 13 \dfrac{\text{m}}{\text{s}}$

4.a. 1st calc

$m = 1,450 \text{ kg}$

$g = 9.80 \dfrac{\text{m}}{\text{s}^2}$

$h = 14.0 \text{ m}$

$E_{G14(at\ h\ =\ 14m)} = ?$

$E_{G14} = mgh = 1,450 \text{ kg} \cdot 9.80 \dfrac{\text{m}}{\text{s}^2} \cdot 14.0 \text{ m} = 199,000 \text{ J}$

$E_{K14(at\ h\ =\ 14m)} = ?$

$E_{G14} + E_{K14} = E_{Gf} + E_{Kf}$

$E_{K14} = E_{Gf} + E_{Kf} - E_{G14} = 0 + 227,360 \text{ J} - 198,940 \text{ J}$

$E_{K14} = 28,400 \text{ J}$

$v_{14} = ?$

$v_{14} = \sqrt{\dfrac{2E_{K14}}{m}} = \sqrt{\dfrac{2 \cdot 28,420 \text{ J}}{1,450 \text{ kg}}} = 6.26 \dfrac{\text{m}}{\text{s}}$

4.a. 2nd calc

$m = 1,450$ kg

$g = 9.80 \dfrac{m}{s^2}$

$h = 12.0$ m

$E_{G12} = ?$

$E_{G12} = mgh = 1,450 \text{ kg} \cdot 9.80 \dfrac{m}{s^2} \cdot 12.0 \text{ m} = 171,000 \text{ J}$

$E_{K12} = ?$

$E_{G12} + E_{K12} = E_{Gf} + E_{Kf}$

$E_{K12} = E_{Gf} + E_{Kf} - E_{G12} = 0 + 227,360 \text{ J} - 170,520 \text{ J}$

$E_{K12} = 56,800 \text{ J}$

$v_{12} = ?$

$v_{12} = \sqrt{\dfrac{2E_{K12}}{m}} = \sqrt{\dfrac{2 \cdot 56,840 \text{ J}}{1,450 \text{ kg}}} = 8.85 \dfrac{m}{s}$

4.a. 3rd calc

$m = 1,450$ kg

$g = 9.80 \dfrac{m}{s^2}$

$h = 10.0$ m

$E_{G10} = ?$

$E_{G10} = mgh = 1,450 \text{ kg} \cdot 9.80 \dfrac{m}{s^2} \cdot 10.0 \text{ m} = 142,000 \text{ J}$

$E_{K10} = ?$

$E_{G10} + E_{K10} = E_{Gf} + E_{Kf}$

$E_{K10} = E_{Gf} + E_{Kf} - E_{G10} = 0 + 227,360 \text{ J} - 142,100 \text{ J}$

$E_{K10} = 85,300 \text{ J}$

$v_{10} = ?$

$v_{10} = \sqrt{\dfrac{2E_{K10}}{m}} = \sqrt{\dfrac{2 \cdot 85,260 \text{ J}}{1,450 \text{ kg}}} = 10.8 \dfrac{m}{s}$

4.a. 4th calc

$m = 1,450$ kg

$g = 9.80 \; \dfrac{\text{m}}{\text{s}^2}$

$h = 8.00$ m

$E_{G8} = ?$

$E_{G8} = mgh = 1,450 \text{ kg} \cdot 9.80 \; \dfrac{\text{m}}{\text{s}^2} \cdot 8.00 \text{ m} = 113,700 \text{ J}$

$E_{K8} = ?$

$E_{G8} + E_{K8} = E_{Gf} + E_{Kf}$

$E_{K8} = E_{Gf} + E_{Kf} - E_{G8} = 0 + 227,360 \text{ J} - 113,700 \text{ J}$

$E_{K8} = 113,700 \text{ J}$

$v_8 = ?$

$v_8 = \sqrt{\dfrac{2E_{K8}}{m}} = \sqrt{\dfrac{2 \cdot 113,700 \text{ J}}{1,450 \text{ kg}}} = 12.5 \; \dfrac{\text{m}}{\text{s}}$

4.a. 5th calc

$m = 1,450$ kg

$g = 9.80 \; \dfrac{\text{m}}{\text{s}^2}$

$h = 6.0$ m

$E_{G6} = ?$

$E_{G6} = mgh = 1,450 \text{ kg} \cdot 9.80 \; \dfrac{\text{m}}{\text{s}^2} \cdot 6.00 \text{ m} = 85,300 \text{ J}$

$E_{K6} = ?$

$E_{G6} + E_{K6} = E_{Gf} + E_{Kf}$

$E_{K6} = E_{Gf} + E_{Kf} - E_{G6} = 0 + 227,360 \text{ J} - 85,260 \text{ J}$

$E_{K6} = 142,00 \text{ J}$

$v_6 = ?$

$v_6 = \sqrt{\dfrac{2E_{K6}}{m}} = \sqrt{\dfrac{2 \cdot 142,100 \text{ J}}{1,450 \text{ kg}}} = 14.0 \; \dfrac{\text{m}}{\text{s}}$

4.a. 6th calc

$m = 1,450 \text{ kg}$

$g = 9.80 \; \dfrac{\text{m}}{\text{s}^2}$

$h = 4.0 \text{ m}$

$E_{G4} = ?$

$E_{G4} = mgh = 1,450 \text{ kg} \cdot 9.80 \; \dfrac{\text{m}}{\text{s}^2} \cdot 4.00 \text{ m} = 56,800 \text{ J}$

$E_{K4} = ?$

$E_{G4} + E_{K4} = E_{Gf} + E_{Kf}$

$E_{K4} = E_{Gf} + E_{Kf} - E_{G4} = 0 + 227,360 \text{ J} - 56,840 \text{ J}$

$E_{K4} = 171,000 \text{ J}$

$v_4 = ?$

$v_4 = \sqrt{\dfrac{2E_{K4}}{m}} = \sqrt{\dfrac{2 \cdot 170,520 \text{ J}}{1,450 \text{ kg}}} = 15.3 \; \dfrac{\text{m}}{\text{s}}$

4.a. 7th calc

$m = 1,450 \text{ kg}$

$g = 9.80 \; \dfrac{\text{m}}{\text{s}^2}$

$h = 2.0 \text{ m}$

$E_{G2} = ?$

$E_{G2} = mgh = 1,450 \text{ kg} \cdot 9.80 \; \dfrac{\text{m}}{\text{s}^2} \cdot 2.00 \text{ m} = 28,400 \text{ J}$

$E_{K2} = ?$

$E_{G2} + E_{K2} = E_{Gf} + E_{Kf}$

$E_{K2} = E_{Gf} + E_{Kf} - E_{G2} = 0 + 227,360 \text{ J} - 28,420 \text{ J}$

$E_{K2} = 199,000 \text{ J}$

$v_2 = ?$

$v_2 = \sqrt{\dfrac{2E_{K2}}{m}} = \sqrt{\dfrac{2 \cdot 198,940 \text{ J}}{1,450 \text{ kg}}} = 16.6 \; \dfrac{\text{m}}{\text{s}}$

Energy Calculations Set 3

1.a.

$$F_w = 27.05 \text{ lb} \cdot \frac{4.45 \text{ N}}{1 \text{ lb}} \cdot 120.37 \text{ N}$$

$$d = 185 \text{ cm} \cdot \frac{1 \text{ m}}{100 \text{ cm}} = 1.85 \text{ m}$$

$$W = Fd = 120.37 \text{ N} \cdot 1.85 \text{ m} = 223 \text{ J}$$

1.b

$$F_w = 120.37 \text{ N}$$

$$h = 1.85 \text{ m}$$

$$g = 9.80 \ \frac{\text{m}}{\text{s}^2}$$

$$m = ?$$

$$F_w = mg$$

$$m = \frac{F_w}{g}$$

$$m = \frac{120.37 \text{ N}}{9.80 \ \dfrac{\text{m}}{\text{s}^2}} = 12.28 \text{ kg}$$

$$E_G = ?$$

$$E_G = mgh = 12.28 \text{ kg} \cdot 9.80 \ \frac{\text{m}}{\text{s}^2} \cdot 1.85 \text{ m} = 223 \text{ J}$$

1.c.

$$m = 12.28 \text{ kg}$$

$$g = 9.80 \ \frac{\text{m}}{\text{s}^2}$$

$$h = 0 \text{ m}$$

$$E_G = ?$$

$$E_G = mgh = 12.28 \text{ kg} \cdot 9.80 \ \frac{\text{m}}{\text{s}^2} \cdot 0 \text{ m} = 0 \text{ J}$$

1.d.

$E_{G_i} = 222.7$ J

$E_{G_f} = 0$ J

$E_{Ki} = 0$ J

$E_{Kf} = ?$

$E_{Gi} + E_{Ki} = E_{G_f} + E_{Kf}$

$E_{Kf} = E_{Gi} + E_{Ki} - E_{G_f} = 222.7$ J $+ 0 - 0 = 223$ J

1.e.

$E_{Kf} = 222.7$ J

$m = 12.28$ kg

$v_f = ?$

$$v_f = \sqrt{\frac{2E_{Kf}}{m}} = \sqrt{\frac{2 \cdot 222.7 \text{ J}}{12.28 \text{ kg}}} = 6.02 \ \frac{\text{m}}{\text{s}}$$

1.f.

$m = 12.28$ kg

$g = 9.80 \ \dfrac{\text{m}}{\text{s}^2}$

$h = \dfrac{1}{2} \cdot 1.85$ m $= 0.925$ m

$E_{G50(50 \text{ percent down})} = mgh = 12.28 \text{ kg} \cdot 9.80 \ \dfrac{\text{m}}{\text{s}^2} \cdot 0.925 \text{ m} = 111$ J

$E_{Kf} = 222.6$ J

$E_{G50} + E_{K50} = E_{Gf} + E_{Kf}$

$E_{K50} = E_{Gf} + E_{Kf} - E_{G50} = 0 + 222.6$ J $- 111.3$ J $= 111$ J

1.g.

$m = 12.28$ kg

$E_{K50} = 111.3$ J

$v_{50} = ?$

$$v_{50} = \sqrt{\frac{2E_{K50}}{m}} = \sqrt{\frac{2 \cdot 111.3 \text{ J}}{12.28 \text{ kg}}} = 4.26 \ \frac{\text{m}}{\text{s}}$$

1.h.

$m = 12.28$ kg

$g = 9.80 \dfrac{m}{s^2}$

$h = 0.10 \cdot 1.85$ m $= 0.185$ m

$E_{G90(90\ percent\ down)} = mgh = 12.28$ kg $\cdot 9.80 \dfrac{m}{s^2} \cdot 0.185$ m $= 22.26$ J

$E_{Kf} = 222.6$ J

$E_{G90} + E_{K90} = E_{Gf} + E_{Kf}$

$E_{K90} = E_{Gf} + E_{Kf} - E_{G90} = 0 + 222.6$ J $- 22.26 = 200.3$ J

$v_{90} = ?$

$v_{90} = \sqrt{\dfrac{2E_{K90}}{m}} = \sqrt{\dfrac{2 \cdot 200.3\ \text{J}}{12.28}} = 5.71 \dfrac{m}{s}$

2.

$d = 197$ ft $\cdot \dfrac{0.3048\ \text{m}}{1\ \text{ft}} = 60.05$ m

$m = 6.016 \times 10^6$ kg

$F_w = mg = 6.016 \times 10^6$ kg $\cdot 9.80 \dfrac{m}{s^2} = 5.896 \times 10^7$ N

$W = ?$

$W = F_w d = 5.896 \times 10^7$ N $\cdot 60.05$ m

$W = 3.54 \times 10^9$ J

3.a.

$m = 5{,}122$ kg

$h_A = 25.0$ m

$h_B = 2.5$ m

$h_C = 18.0$ m

$v_B = ?$

$E_{GA} = mgh_A = 5{,}122 \text{ kg} \cdot 9.80 \ \dfrac{\text{m}}{\text{s}^2} \cdot 25.0 \text{ m} = 1.2549 \times 10^6 \text{ J}$

$E_{GB} = mgh_B = 5{,}122 \text{ kg} \cdot 9.80 \ \dfrac{\text{m}}{\text{s}^2} \cdot 2.5 \text{ m} = 1.2549 \times 10^5 \text{ J}$

$E_{KA} = 0 \text{ J}$

$E_{GA} + E_{KA} = E_{GB} + E_{KB}$

$E_{KB} = E_{GA} + E_{KA} - E_{GB} = 1.2549 \times 10^6 \text{ J} + 0 \text{ J} - 1.2549 \times 10^5 \text{ J}$

$E_{KB} = 1.1294 \times 10^6 \text{ J}$

$v_B = \sqrt{\dfrac{2 E_{KB}}{m}} = \sqrt{\dfrac{2 \cdot 1.1294 \times 10^6 \text{ J}}{5{,}122 \text{ kg}}} = 21.0 \ \dfrac{\text{m}}{\text{s}}$

3.b.

$m = 5{,}122$ kg

$h_A = 25.0$ m

$h_C = 18.0$ m

$v_C = ?$

$E_{GA} = 1.2549 \times 10^6 \text{ J}$

$E_{GC} = mgh_C = 5{,}122 \text{ kg} \cdot 9.80 \ \dfrac{\text{m}}{\text{s}^2} \cdot 18.0 \text{ m} = 9.0352 \times 10^5 \text{ J}$

$E_{KA} = 0 \text{ J}$

$E_{GA} + E_{KA} = E_{GC} + E_{KC}$

$E_{KC} = E_{GA} + E_{KA} - E_{GC} = 1.2549 \times 10^6 \text{ J} + 0 \text{ J} - 9.0352 \times 10^5 \text{ J}$

$E_{KC} = 3.5138 \times 10^5 \text{ J}$

$v_C = \sqrt{\dfrac{2 E_{KC}}{m}} = \sqrt{\dfrac{2 \cdot 3.5138 \times 10^5 \text{ J}}{5{,}122 \text{ kg}}} = 11.7 \ \dfrac{\text{m}}{\text{s}}$

4.a.

$$F_w = 104.6 \text{ lb} \cdot \frac{4.45 \text{ N}}{1 \text{ lb}} = 4.6547 \times 10^2 \text{ N}$$

$$d = 13 \text{ steps} \cdot \frac{16.5 \text{ cm}}{1 \text{ step}} \cdot \frac{1 \text{ m}}{100 \text{ cm}} = 2.145 \text{ m}$$

$$W = ?$$

$$W = F_w d = 4.6547 \times 10^2 \text{ N} \cdot 2.145 \text{ m} = 998 \text{ J}$$

4.b.

$$F_w = 4.6547 \times 10^2 \text{ N}$$

$$F_w = mg$$

$$m = \frac{F_w}{g} = \frac{4.6547 \times 10^2 \text{ N}}{9.80 \frac{\text{m}}{\text{s}^2}} = 47.497 \text{ kg}$$

$$h_i = 2.145 \text{ m}$$

$$v_f = ?$$

$$E_{Gi} = mgh_i = 47.497 \text{ kg} \cdot 9.80 \frac{\text{m}}{\text{s}^2} \cdot 2.145 \text{ m}$$

$$E_{Gi} = 998.43 \text{ J}$$

$$E_{Gi} + E_{Ki} = E_{Gf} + E_{Kf}$$

$$E_{Kf} = E_{Gi} + E_{Ki} - E_{Gf} = 998.43 \text{ J} + 0 - 0$$

$$E_{Kf} = 998.43 \text{ J}$$

$$v_f = \sqrt{\frac{2E_{Kf}}{m}} = \sqrt{\frac{2 \cdot 998.43 \text{ J}}{47.497 \text{ kg}}} = 6.48 \frac{\text{m}}{\text{s}}$$

5.

$$m = 351 \text{ g} \cdot \frac{1 \text{ kg}}{1000 \text{ g}} = 0.351 \text{ kg}$$

$$v_i = 500.00 \; \frac{\text{cm}}{\text{s}} \cdot \frac{1 \text{ m}}{100 \text{ cm}} = 5.00 \; \frac{\text{m}}{\text{s}}$$

$$v_f = 0 \; \frac{\text{m}}{\text{s}}$$

$$h_i = 0 \text{ m}$$

$$h_f = ?$$

$$E_{Ki} = \frac{1}{2} m v^2 = \frac{1}{2} \cdot 0.351 \text{ kg} \cdot (5.00 \; \frac{\text{m}}{\text{s}})^2$$

$$E_{Ki} = 4.388 \text{ J}$$

$$E_{Gi} + E_{Ki} = E_{Gf} + E_{Kf}$$

$$E_{Gf} = E_{Gi} + E_{Ki} - E_{Kf} = 0 \text{ J} + 4.388 \text{ J} - 0 \text{ J}$$

$$E_{Gf} = 4.388 \text{ J}$$

$$E_{Gf} = mgh_f$$

$$h_f = \frac{E_{Gf}}{mg} = \frac{4.388 \text{ J}}{0.351 \text{ kg} \cdot 9.80 \; \frac{\text{m}}{\text{s}^2}} = 1.28 \text{ m}$$

6.

$$F_w = 4,294 \text{ lb} \cdot \frac{4.45 \text{ N}}{1 \text{ lb}} = 1.9108 \times 10^4 \text{ N}$$

$$F_w = mg$$

$$m = \frac{F_w}{g} = \frac{1.9108 \times 10^4 \text{ N}}{9.80 \dfrac{\text{m}}{\text{s}^2}} = 1.9498 \times 10^3 \text{ kg}$$

$$v_i = 27.89 \frac{\text{ft}}{\text{s}} \cdot \frac{0.3048 \text{ m}}{1 \text{ ft}} = 8.5009 \frac{\text{m}}{\text{s}}$$

$$h_f = 7.710 \text{ ft} \cdot \frac{0.3048 \text{ m}}{1 \text{ ft}} = 2.3500 \text{ m}$$

$$v_f = ?$$

$$E_{Ki} = \frac{1}{2} m v_i^2$$

$$E_{Ki} = \frac{1}{2} \cdot 1.9498 \times 10^3 \text{ kg} \cdot (8.5009 \frac{\text{m}}{\text{s}})^2 = 7.0451 \times 10^4 \text{ J}$$

$$E_{Gf} = mgh = 1.9498 \times 10^3 \text{ kg} \cdot 9.80 \frac{\text{m}}{\text{s}^2} \cdot 2.3500 \text{ m}$$

$$E_{Gf} = 4.4904 \times 10^4 \text{ J}$$

$$E_{Gi} + E_{Ki} = E_{Gf} + E_{Kf}$$

$$E_{Kf} = E_{Gi} + E_{Ki} - E_{Gf} = 0 \text{ J} + 7.0451 \times 10^4 \text{ J} - 4.4904 \times 10^4 \text{ J}$$

$$E_{Kf} = 2.5547 \times 10^4 \text{ J}$$

$$v_f = \sqrt{\frac{2 E_{Kf}}{m}} = \sqrt{\frac{2 \cdot 2.5547 \times 10^4 \text{ J}}{1.9498 \times 10^3 \text{ kg}}} = 5.12 \frac{\text{m}}{\text{s}}$$

7.

$h_f = 6.500$ m

$m = 950.0$ g $\cdot \dfrac{1 \text{ kg}}{1000 \text{ g}} = 0.9500$ kg

$v_f = 1.000 \; \dfrac{\text{m}}{\text{s}}$

$v_i = ?$

$E_{Gi} = 0$ J

$E_{Gf} = mgh = 0.9500 \text{ kg} \cdot 9.803 \; \dfrac{\text{m}}{\text{s}^2} \cdot 6.500 \text{ m} = 60.534$ J

$E_{Kf} = \dfrac{1}{2} m v_f^{\,2} = \dfrac{1}{2} \cdot 0.9500 \text{ kg} \cdot (1.000 \; \dfrac{\text{m}}{\text{s}})^2 = 0.4750$ J

$E_{Gi} + E_{Ki} = E_{Gf} + E_{Kf}$

$E_{Ki} = E_{Gf} + E_{Kf} - E_{Gi} = 60.534 \text{ J} + 0.4750 \text{ J} - 0 \text{ J}$

$E_{Ki} = 61.009$ J

$v_i = \sqrt{\dfrac{2 E_{Ki}}{m}} = \sqrt{\dfrac{2 \cdot 61.009 \text{ J}}{0.9500 \text{ kg}}} = 11.33 \; \dfrac{\text{m}}{\text{s}}$

Energy Calculations Set 4

1.

$$F_w = 3,420.1 \text{ lb} \cdot \frac{4.45 \text{ N}}{1 \text{ lb}} = 1.52194 \times 10^4 \text{ N}$$

$$h_i = 31.0 \text{ ft} \cdot \frac{0.3048 \text{ m}}{1 \text{ ft}} = 9.449 \text{ m}$$

$$h_f = 6.50 \text{ ft} \cdot \frac{0.3048 \text{ m}}{1 \text{ ft}} = 1.981 \text{ m}$$

$$v_f = ?$$

$$F_w = mg$$

$$m = \frac{F_w}{g} = \frac{1.52194 \times 10^4 \text{ N}}{9.80 \; \frac{\text{m}}{\text{s}^2}} = 1.5530 \times 10^3 \text{ kg}$$

$$E_{Gi} = mgh_i = 1.5530 \times 10^3 \text{ kg} \cdot 9.80 \; \frac{\text{m}}{\text{s}^2} \cdot 9.449 \text{ m} = 1.4381 \times 10^5 \text{ J}$$

$$E_{Gf} = mgh_f = 1.5530 \times 10^3 \text{ kg} \cdot 9.80 \; \frac{\text{m}}{\text{s}^2} \cdot 1.981 \text{ m} = 3.0150 \times 10^4 \text{ J}$$

$$E_{Ki} = 0 \text{ J}$$

$$E_{Gi} + E_{Ki} = E_{Gf} + E_{Kf}$$

$$E_{Kf} = E_{Gi} + E_{Ki} - E_{Gf} = 1.4381 \times 10^5 \text{ J} + 0 \text{ J} - 3.0150 \times 10^4 \text{ J}$$

$$E_{Kf} = 1.1366 \times 10^5 \text{ J}$$

$$v_f = \sqrt{\frac{2E_{Kf}}{m}} = \sqrt{\frac{2 \cdot 1.1366 \times 10^5 \text{ J}}{1.5530 \times 10^3 \text{ kg}}} = 12.1 \; \frac{\text{m}}{\text{s}}$$

2.

$m = 2,200 \text{ kg}$

$h_f = 33.87 \text{ m}$

$h_i = 37.00 \text{ m}$

$v_f = 17.88 \dfrac{\text{m}}{\text{s}}$

$v_i = ?$

$E_{Gi} = mgh_i = 2,200 \text{ kg} \cdot 9.80 \dfrac{\text{m}}{\text{s}^2} \cdot 37.00 \text{ m} = 7.98 \times 10^5 \text{ J}$

$E_{Gf} = mgh_f = 2,200 \text{ kg} \cdot 9.80 \dfrac{\text{m}}{\text{s}^2} \cdot 33.87 \text{ m} = 7.31 \times 10^5 \text{ J}$

$E_{Kf} = \dfrac{1}{2}mv^2 = \dfrac{1}{2} \cdot 2,200 \text{ kg} \cdot (17.88 \dfrac{\text{m}}{\text{s}})^2 = 3.52 \times 10^5 \text{ J}$

$E_{Gi} + E_{Ki} = E_{Gf} + E_{Kf}$

$E_{Ki} = E_{Gf} + E_{Kf} - E_{Gi}$

$E_{Ki} = 7.31 \times 10^5 \text{ J} + 3.52 \times 10^5 \text{ J} - 7.98 \times 10^5 \text{ J} = 2.85 \times 10^5 \text{ J}$

$v_i = \sqrt{\dfrac{2E_{Ki}}{m}} = \sqrt{\dfrac{2 \cdot 2.85 \times 10^5 \text{ J}}{2,200 \text{ kg}}} = 16 \dfrac{\text{m}}{\text{s}}$

3.

$$m = 1.873 \times 10^{-2} \text{ mg} \cdot \frac{1 \text{ g}}{1000 \text{ mg}} \cdot \frac{1 \text{ kg}}{1000 \text{ g}} = 1.873 \times 10^{-8} \text{ kg}$$

$$h_i = 2.177 \text{ cm} \cdot \frac{1 \text{ m}}{100 \text{ cm}} = 0.02177 \text{ m}$$

$$v_i = 202.75 \frac{\text{cm}}{\text{s}} \cdot \frac{1 \text{ m}}{100 \text{ cm}} = 2.0275 \frac{\text{m}}{\text{s}}$$

$$h_f = 7.50 \text{ cm} \cdot \frac{1 \text{ m}}{100 \text{ cm}} = 0.0750 \text{ m}$$

$$v_f = ?$$

$$E_{Gi} = mgh_i = 1.873 \times 10^{-8} \text{ kg} \cdot 9.80 \frac{\text{m}}{\text{s}^2} \cdot 0.02177 \text{ m} = 3.996 \times 10^{-9} \text{ J}$$

$$E_{Ki} = \frac{1}{2} mv^2 = \frac{1}{2} \cdot 1.873 \times 10^{-8} \text{ kg} \cdot (2.0275 \frac{\text{m}}{\text{s}})^2 = 3.850 \times 10^{-8} \text{ J}$$

$$E_{Gf} = mgh_f = 1.873 \times 10^{-8} \text{ kg} \cdot 9.80 \frac{\text{m}}{\text{s}^2} \cdot 0.0750 \text{ m} = 1.377 \times 10^{-8} \text{ J}$$

$$E_{Gi} + E_{Ki} = E_{Gf} + E_{Kf}$$

$$E_{Kf} = E_{Gi} + E_{Ki} - E_{Gf} = 3.996 \times 10^{-9} \text{ J} + 3.850 \times 10^{-8} \text{ J} - 1.377 \times 10^{-8} \text{ J}$$

$$E_{Kf} = 2.873 \times 10^{-8} \text{ J}$$

$$v_f = \sqrt{\frac{2E_{Kf}}{m}} = \sqrt{\frac{2 \cdot 2.873 \times 10^{-8} \text{ J}}{0.0750 \text{ m}}} = 1.75 \frac{\text{m}}{\text{s}}$$

4.

$$m = 9.0022 \ \mu g \cdot \frac{1 \ g}{10^6 \ \mu g} \cdot \frac{1 \ kg}{1000 \ g} = 9.0022 \times 10^{-9} \ kg$$

$$h_i = 9.0125 \ cm \cdot \frac{1 \ m}{100 \ cm} = 0.090125 \ m$$

$$v_i = 0$$

$$v_f = 85.160 \ \frac{cm}{s} \cdot \frac{1 \ m}{100 \ cm} = 0.85160 \ \frac{m}{s}$$

$$h_f = ?$$

$$E_{Ki} + E_{Gi} = E_{Kf} + E_{Gf}$$

$$E_{Gf} = E_{Ki} + E_{Gi} - E_{Kf}$$

$$E_{Ki} = 0$$

$$E_{Gi} = mgh_i = 9.0022 \times 10^{-9} \ kg \cdot 9.80 \ \frac{m}{s^2} \cdot 0.090125 \ m = 7.951 \times 10^{-9} \ J$$

$$E_{Kf} = \frac{1}{2} m v_f^2 = \frac{1}{2} \cdot 9.0022 \times 10^{-9} \ kg \cdot \left(0.85160 \ \frac{m}{s} \right)^2 = 3.264 \times 10^{-9} \ J$$

$$E_{Gf} = 0 + 7.951 \times 10^{-9} \ J - 3.264 \times 10^{-9} \ J = 4.687 \times 10^{-9} \ J$$

$$E_{Gf} = mgh_f$$

$$h_f = \frac{E_{Gf}}{mg} = \frac{4.687 \times 10^{-9} \ J}{9.0022 \times 10^{-9} \ kg \cdot 9.80 \ \frac{m}{s^2}} = 0.0531 \ m \cdot \frac{100 \ cm}{1 \ m} = 5.31 \ cm$$

5.

$$v_i = 1.562 \ \frac{\text{ft}}{\text{s}} \cdot \frac{0.3048 \text{ m}}{1 \text{ ft}} = 0.40761 \ \frac{\text{m}}{\text{s}}$$

$$F_w = 4.1843 \text{ lb} \cdot \frac{4.45 \text{ N}}{1 \text{ lb}} = 18.6201 \text{ N}$$

$$v_f = 24.75 \ \frac{\text{ft}}{\text{s}} \cdot \frac{0.3048 \text{ m}}{1 \text{ ft}} = 7.5438 \ \frac{\text{m}}{\text{s}}$$

$$h_f = 0 \text{ m}$$

$$h_i = ?$$

$$F_w = mg$$

$$m = \frac{F_w}{g} = \frac{18.6201 \text{ N}}{9.80 \ \frac{\text{m}}{\text{s}^2}} = 1.900 \text{ kg}$$

$$E_{Ki} = \frac{1}{2}mv_i^2 = \frac{1}{2} \cdot 1.900 \text{ kg} \cdot (0.4761 \ \frac{\text{m}}{\text{s}})^2 = 0.2153 \text{ J}$$

$$E_{Gf} = 0 \text{ J}$$

$$E_{Kf} = \frac{1}{2}mv_f^2 = \frac{1}{2} \cdot 1.900 \text{ kg} \cdot (7.5438 \ \frac{\text{m}}{\text{s}})^2 = 54.063 \text{ J}$$

$$E_{Gi} + E_{Ki} = E_{Gf} + E_{Kf}$$

$$E_{Gi} = E_{Gf} + E_{Kf} - E_{Ki} = 0 \text{ J} + 54.063 \text{ J} - 0.2153 \text{ J}$$

$$E_{Gi} = 53.85 \text{ J}$$

$$E_{Gi} = mgh_i$$

$$h_i = \frac{E_{Gi}}{mg} = \frac{53.85 \text{ J}}{1.900 \text{ kg} \cdot 9.80 \ \frac{\text{m}}{\text{s}^2}} = 2.89 \text{ m}$$

Chapter 6

Temperature Unit Conversions

1.a.

$T_C = 32.0°C$

$T_F = ?$

$T_C = \dfrac{5}{9}(T_F - 32°)$

$T_F = \dfrac{9}{5}T_C + 32° = \dfrac{9}{5} \cdot 32.0°C + 32° = 89.6°F$

1.b.

$T_C = 32.0°C$

$T_K = ?$

$T_K = T_C + 273.2$

$T_K = 32.0°C + 273.2 = 305.2 \text{ K}$

2.a.

$T_F = 56.5°F$

$T_C = ?$

$T_C = \dfrac{5}{9}(T_F - 32°)$

$T_C = \dfrac{5}{9}(56.5°F - 32°) = 13.6°C$

2.b.

$T_F = 56.5°F$

$T_C = 13.6°C$

$T_K = ?$

$T_K = T_C + 273.2$

$T_K = 13.6°C + 273.2 = 286.8 \text{ K}$

3.a.

$T_K = 455.0$ K

$T_C = 181.8°C$

$T_F = ?$

$T_C = \dfrac{5}{9}(T_F - 32°)$

$T_F = \dfrac{9}{5}T_C + 32°$

$T_F = \dfrac{9}{5} \cdot 181.8°C + 32° = 359.2°F$

3.b.

$T_K = 455.0$ K

$T_C = ?$

$T_K = T_C + 273.2$

$T_C = T_K - 273.2 = 455.0 \text{ K} - 273.2 = 181.8°C$

4.a.

$T_F = -17.9°F$

$T_C = ?$

$T_C = \dfrac{5}{9}(T_F - 32°)$

$T_C = \dfrac{5}{9}(-17.9°F - 32°) = -27.7°C$

4.b.

$T_F = -17.9°F$

$T_C = -27.7°C$

$T_K = ?$

$T_K = T_C + 273.2$

$T_K = -27.7°C + 273.2 = 245.5$ K

5.a.

$T_C = -41.6°C$

$T_F = ?$

$T_C = \dfrac{5}{9}(T_F - 32°)$

$T_F = \dfrac{9}{5}T_C + 32°$

$T_F = \left(\dfrac{9}{5}\cdot(-41.6°C)\right) + 32° = -42.9°F$

5.b.

$T_C = -41.6°C$

$T_K = ?$

$T_K = T_C + 273.2$

$T_K = -41.6°C + 273.2 = 231.6 \text{ K}$

6.a.

$T_K = 79.0 \text{ K}$

$T_C = ?$

$T_K = T_C + 273.2$

$T_C = T_K - 273.2$

$T_C = 79.0 \text{ K} - 273.2 = -194.2°C$

6.b.

$T_K = 79.0 \text{ K}$

$T_C = -194.2°C$

$T_F = ?$

$T_C = \dfrac{5}{9}(T_F - 32°)$

$T_F = \dfrac{9}{5}T_C + 32°$

$T_F = \left(\dfrac{9}{5}\cdot(-194.2°C)\right) + 32° = -317.6° \text{ F}$

Chapter 7

1.

$f = 60.0$ Hz

$\tau = ?$

$\tau = \dfrac{1}{f}$

$\tau = \dfrac{1}{60.0 \text{ Hz}} = 0.0167$ s

2.

$\tau = 2.155 \times 10^{-5}$ s

$f = ?$

$\tau = \dfrac{1}{f}$

$f = \dfrac{1}{\tau} = \dfrac{1}{2.155 \times 10^{-5} \text{ s}} = 46.40$ kHz

---

3.

$f = 26.0 \text{ kHz} \cdot \dfrac{1000 \text{ Hz}}{1 \text{ kHz}} = 2.60 \times 10^4$ Hz

$\tau = ?$

$\tau = \dfrac{1}{f} = \dfrac{1}{2.60 \times 10^4 \text{ Hz}} = 3.85 \times 10^{-5}$ s

---

4.

$f = 2.60 \times 10^4$ Hz

$v = 342 \ \dfrac{\text{m}}{\text{s}}$

$\lambda = ?$

$v = \lambda f$

$\lambda = \dfrac{v}{f} = \dfrac{342 \ \dfrac{\text{m}}{\text{s}}}{2.60 \times 10^4 \text{ Hz}} = 1.32 \times 10^{-2} \text{ m} \cdot \dfrac{100 \text{ cm}}{1 \text{ m}} = 1.32$ cm

---

5.

$$f = 89.5 \text{ MHz} \cdot \frac{10^6 \text{ Hz}}{1 \text{ MHz}} = 8.95 \times 10^7 \text{ Hz}$$

$$v = c = 3.00 \times 10^8 \ \frac{\text{m}}{\text{s}}$$

$$\lambda = ?$$

$$\tau = ?$$

$$v = \lambda f$$

$$\lambda = \frac{v}{f} = \frac{3.00 \times 10^8 \ \frac{\text{m}}{\text{s}}}{8.95 \times 10^7 \text{ Hz}} = 3.35 \text{ m}$$

$$\tau = \frac{1}{f} = \frac{1}{8.95 \times 10^7 \text{ Hz}} = 1.12 \times 10^{-8} \text{ s} \cdot \frac{10^6 \ \mu\text{s}}{1 \text{ s}} = 0.0112 \ \mu\text{s}$$

6.

$$v = c = 3.00 \times 10^8 \ \frac{\text{m}}{\text{s}}$$

$$f = 1{,}310 \text{ kHz} \cdot \frac{1000 \text{ Hz}}{1 \text{ kHz}} = 1.31 \times 10^6 \text{ Hz}$$

$$\lambda = ?$$

$$\tau = ?$$

$$v = \lambda f$$

$$\lambda = \frac{v}{f} = \frac{3.00 \times 10^8 \ \frac{\text{m}}{\text{s}}}{1.31 \times 10^6 \text{ Hz}} = 229 \text{ m}$$

$$\tau = \frac{1}{f} = \frac{1}{1.31 \times 10^6 \text{ Hz}} = 7.64 \times 10^{-7} \text{ s} \cdot \frac{10^6 \ \mu\text{s}}{1 \text{ s}} = 0.763 \ \mu\text{s}$$

8.

$$v = c = 3.00 \times 10^8 \ \frac{\text{m}}{\text{s}}$$

$$\lambda = 542 \ \text{nm} \cdot \frac{1 \ \text{m}}{10^9 \ \text{nm}} = 5.42 \times 10^{-7} \ \text{m}$$

$$\tau = ?$$

$$f = ?$$

$$v = \lambda f$$

$$f = \frac{v}{\lambda} = \frac{3.00 \times 10^8 \ \frac{\text{m}}{\text{s}}}{5.42 \times 10^{-7} \ \text{m}} = 5.54 \times 10^{14} \ \text{Hz} \cdot \frac{1 \ \text{GHz}}{10^9 \ \text{Hz}} = 5.54 \times 10^5 \ \text{GHz}$$

$$\tau = \frac{1}{f} = \frac{1}{5.54 \times 10^{14} \ \text{Hz}} = 1.81 \times 10^{-15} \ \text{s} \cdot \frac{10^9 \ \text{ns}}{1 \ \text{s}} = 1.81 \times 10^{-6} \ \text{ns}$$

9.

$$\lambda = 10.6 \ \mu\text{m} \cdot \frac{1 \ \text{m}}{10^6 \ \mu\text{m}} = 1.06 \times 10^{-5} \ \text{m}$$

$$v = 3.00 \times 10^8 \ \frac{\text{m}}{\text{s}}$$

$$\tau = ?$$

$$f = ?$$

$$v = \lambda f$$

$$f = \frac{v}{\lambda} = \frac{3.00 \times 10^8 \ \frac{\text{m}}{\text{s}}}{1.06 \times 10^{-5} \ \text{m}} = 2.83 \times 10^{13} \ \text{Hz} \cdot \frac{1 \ \text{MHz}}{10^6 \ \text{Hz}} = 2.83 \times 10^7 \ \text{MHz}$$

$$\tau = \frac{1}{f} = \frac{1}{2.83 \times 10^{13} \ \text{Hz}} = 3.53 \times 10^{-14} \ \text{s} \cdot \frac{10^6 \ \mu\text{s}}{1 \ \text{s}} = 3.53 \times 10^{-8} \ \mu\text{s}$$

10.

$$f = 33 \text{ kHz} \cdot \frac{1000 \text{ Hz}}{1 \text{ kHz}} = 3.3 \times 10^4 \text{ Hz}$$

$$v = 342 \ \frac{\text{m}}{\text{s}}$$

$$\tau = ?$$

$$\lambda = ?$$

$$\tau = \frac{1}{f} = \frac{1}{3.3 \times 10^4 \text{ Hz}} = 3.03 \times 10^{-5} \text{ s} \cdot \frac{10^3 \text{ ms}}{1 \text{ s}} = 0.030 \text{ ms}$$

$$v = \lambda f$$

$$\lambda = \frac{v}{f} = \frac{342 \ \frac{\text{m}}{\text{s}}}{3.3 \times 10^4 \text{ Hz}} = 1.04 \times 10^{-2} \text{ m} \cdot \frac{1000 \text{ mm}}{1 \text{ m}} = 1.0 \times 10^1 \text{ mm}$$

11.

$$v = 3.00 \times 10^8 \ \frac{\text{m}}{\text{s}}$$

$$\lambda = 2.00 \text{ mm} \cdot \frac{1 \text{ m}}{10^3 \text{ mm}} = 2.00 \times 10^{-3} \text{ m}$$

$$v = \lambda f$$

$$f = \frac{v}{\lambda} = \frac{3.00 \times 10^8 \ \frac{\text{m}}{\text{s}}}{2.00 \times 10^{-3} \text{ m}} = 1.50 \times 10^{11} \text{ Hz}$$

12.

$$\lambda_b = 470 \text{ nm} \cdot \frac{1 \text{ m}}{10^9 \text{ nm}} = 4.7 \times 10^{-7} \text{ m}$$

$$\lambda_g = 550 \text{ nm} \cdot \frac{1 \text{ m}}{10^9 \text{ nm}} = 5.5 \times 10^{-7} \text{ m}$$

$$\lambda_r = 680 \text{ nm} \cdot \frac{1 \text{ m}}{10^9 \text{ nm}} = 6.8 \times 10^{-7} \text{ m}$$

$$v = 3.00 \times 10^8 \ \frac{\text{m}}{\text{s}}$$

$$f_b = ?$$
$$f_g = ?$$
$$f_r = ?$$
$$v = \lambda f$$

$$f = \frac{v}{\lambda}$$

$$f_b = \frac{3.00 \times 10^8 \ \dfrac{\text{m}}{\text{s}}}{4.7 \times 10^{-7} \text{ m}} = 6.4 \times 10^{14} \text{ Hz}$$

$$f_g = \frac{3.00 \times 10^8 \ \dfrac{\text{m}}{\text{s}}}{5.5 \times 10^{-7} \text{ m}} = 5.5 \times 10^{14} \text{ Hz}$$

$$f_r = \frac{3.00 \times 10^8 \ \dfrac{\text{m}}{\text{s}}}{6.8 \times 10^{-7} \text{ m}} = 4.4 \times 10^{14} \text{ Hz}$$

13.

$$f = 20.00 \text{ Hz}$$

$$v = 342.0 \ \frac{\text{m}}{\text{s}}$$

$$\lambda = ?$$
$$v = \lambda f$$

$$\lambda = \frac{v}{f} = \frac{342.0 \ \dfrac{\text{m}}{\text{s}}}{20.00 \text{ Hz}} = 17.10 \text{ m}$$

14.a.

$$f = 1.00 \text{ kHz} \cdot \frac{1000 \text{ Hz}}{1 \text{ kHz}} = 1.00 \times 10^3 \text{ Hzs}$$

$$v = 342 \ \frac{m}{s}$$

$$\lambda = ?$$

$$v = \lambda f$$

$$\lambda = \frac{v}{f} = \frac{342 \ \frac{m}{s}}{1.00 \times 10^3 \text{ Hz}} = 0.342 \text{ m}$$

14.b.

$$f = 1.00 \text{ kHz} \cdot \frac{1000 \text{ Hz}}{1 \text{ kHz}} = 1.00 \times 10^3 \text{ Hz}$$

$$v = 1,402 \ \frac{m}{s}$$

$$\lambda = ?$$

$$v = \lambda f$$

$$\lambda = \frac{v}{f} = \frac{1,402 \ \frac{m}{s}}{1.00 \times 10^3 \text{ Hz}} = 1.40 \text{ m}$$

14.c.

$$f = 1.00 \text{ kHz} \cdot \frac{1000 \text{ Hz}}{1 \text{ kHz}} = 1.00 \times 10^3 \text{ Hz}$$

$$v = 5,130 \ \frac{m}{s}$$

$$\lambda = ?$$

$$v = \lambda f$$

$$\lambda = \frac{v}{f} = \frac{5,130 \ \frac{m}{s}}{1.00 \times 10^3 \text{ Hz}} = 5.13 \text{ m}$$

14.d.

$$f = 1.00 \text{ kHz} \cdot \frac{1000 \text{ Hz}}{1 \text{ kHz}} = 1.00 \times 10^3 \text{ Hz}$$

$$v = 965 \ \frac{\text{m}}{\text{s}}$$

$$\lambda = ?$$

$$v = \lambda f$$

$$\lambda = \frac{v}{f} = \frac{965 \ \frac{\text{m}}{\text{s}}}{1.00 \times 10^3 \text{ Hz}} = 0.965 \text{ m}$$

15.a.

$$f = 4.67 \times 10^{20} \text{ Hz}$$

$$v = 3.00 \times 10^8 \ \frac{\text{m}}{\text{s}}$$

$$\lambda = ?$$

$$v = \lambda f$$

$$\lambda = \frac{v}{f} = \frac{3.00 \times 10^8 \ \frac{\text{m}}{\text{s}}}{4.67 \times 10^{20} \text{ Hz}} = 6.42 \times 10^{-13} \text{ m}$$

$$6.42 \times 10^{-13} \text{ m} \cdot \frac{10^9 \text{ nm}}{1 \text{ m}} = 0.000642 \text{ nm}$$

15.b.

$$f = 9.9876 \times 10^{18} \text{ Hz}$$

$$v = 3.00 \times 10^8 \ \frac{\text{m}}{\text{s}}$$

$$\lambda = ?$$

$$v = \lambda f$$

$$\lambda = \frac{v}{f} = \frac{3.00 \times 10^8 \ \frac{\text{m}}{\text{s}}}{9.9876 \times 10^{18} \text{ Hz}} = 3.00 \times 10^{-11} \text{ m}$$

$$3.00 \times 10^{-11} \text{ m} \cdot \frac{10^9 \text{ nm}}{1 \text{ m}} = 0.0300 \text{ nm}$$

15.c.

$$f = 2.555 \times 10^{10} \text{ Hz}$$

$$v = 3.00 \times 10^{8} \ \frac{\text{m}}{\text{s}}$$

$$\lambda = ?$$

$$v = \lambda f$$

$$\lambda = \frac{v}{f} = \frac{3.00 \times 10^{8} \ \frac{\text{m}}{\text{s}}}{2.555 \times 10^{10} \text{ Hz}} = 1.17 \times 10^{-2} \text{ m}$$

$$1.17 \times 10^{-2} \text{ m} \cdot \frac{10^{9} \text{ nm}}{1 \text{ m}} = 11,700,000 \text{ nm}$$

15.d.

$$f = 1.172 \times 10^{15} \text{ Hz}$$

$$v = 3.00 \times 10^{8} \ \frac{\text{m}}{\text{s}}$$

$$\lambda = ?$$

$$v = \lambda f$$

$$\lambda = \frac{v}{f} = \frac{3.00 \times 10^{8} \ \frac{\text{m}}{\text{s}}}{1.172 \times 10^{15} \text{ Hz}} = 2.56 \times 10^{-7} \text{ m}$$

$$2.56 \times 10^{-7} \text{ m} \cdot \frac{10^{9} \text{ nm}}{1 \text{ m}} = 256 \text{ nm}$$

15.e.

$$f = 2.83 \times 10^{13} \text{ Hz}$$

$$v = 3.00 \times 10^{8} \ \frac{\text{m}}{\text{s}}$$

$$\lambda = ?$$

$$v = \lambda f$$

$$\lambda = \frac{v}{f} = \frac{3.00 \times 10^{8} \ \frac{\text{m}}{\text{s}}}{2.83 \times 10^{13} \text{ Hz}} = 1.06 \times 10^{-5} \text{ m}$$

$$1.06 \times 10^{-5} \text{ m} \cdot \frac{10^{9} \text{ nm}}{1 \text{ m}} = 10,600 \text{ nm}$$

16.

$\lambda = 1.17 \times 10^{-2}$ m

$D = 3 \text{ mm} \cdot \dfrac{1 \text{ m}}{10^3 \text{ mm}} = 0.003$ m

$ratio = ?$

$ratio = \dfrac{\lambda}{D} = \dfrac{1.17 \times 10^{-2} \text{ m}}{0.003 \text{ m}} = 4$

Chapter 8

Introductory Circuit Calculations

1.

$I = 13.00$ A

$V = 25.00$ V

$R = ?$

$V = IR$

$R = \dfrac{V}{I} = \dfrac{25.00 \text{ V}}{13.00 \text{ A}} = 1.923 \ \Omega$

2.

$V = 24$ V

$R = 250 \ \Omega$

$I = ?$

$V = IR$

$I = \dfrac{V}{R} = \dfrac{24 \text{ V}}{250 \ \Omega} = 0.096 \text{ A} \cdot \dfrac{1000 \text{ mA}}{1 \text{ A}} = 96 \text{ mA}$

3.

$R = 12.20$ kΩ

$V = 4.500$ V

$I = ?$

$V = IR$

$I = \dfrac{V}{R} = \dfrac{4.500 \text{ V}}{12.20 \text{ k}\Omega} = 0.3689 \text{ mA}$

4.

$I = 0.0300$ mA

$R = 33.3 \text{ M}\Omega \ \cdot \dfrac{10^3 \text{ k}\Omega}{1 \text{ M}\Omega} = 3.33 \times 10^4 \text{ k}\Omega$

$V = ?$

$V = IR = 0.0300 \text{ mA} \cdot 3.33 \times 10^4 \text{ k}\Omega \ = 999 \text{ V}$

5.a.

$I = 13.00$ A

$V = 25.00$ V

$P = ?$

$P = VI = 25.00$ V $\cdot 13.00$ A $= 325.0$ W

5.b.

$V = 24$ V

$I = 96$ mA

$P = ?$

$P = VI = 24$ V $\cdot 96$ mA $= 2{,}300$ mW $= 2.3$ W

5.c.

$V = 4.500$ V

$I = 0.3689$ mA

$P = ?$

$P = VI = 4.500$ V $\cdot 0.3689$ mA $= 1.660$ mW

6.

$V = 120$ V

$P = 60.00$ W

$R = ?$

$I = ?$

$P = VI$

$$I = \frac{P}{V} = \frac{60.00 \text{ W}}{120 \text{ V}} = 0.50 \text{ A}$$

$V = IR$

$$R = \frac{V}{I} = \frac{120 \text{ V}}{0.50 \text{ A}} = 240 \text{ } \Omega$$

7.

$V = 120$ V

$I = 12$ A

$P = ?$

$$P = VI = 120 \text{ V} \cdot 12 \text{ A} = 1{,}440 \text{ W} \cdot \frac{1 \text{ kW}}{1000 \text{ W}} = 1.4 \text{ kW}$$

8.

$I = 13.5\ \mu A$

$V = 6.0\ V$

$P = ?$

$P = VI = 6.0\ V \cdot 13.5\ \mu A = 81\ \mu W$

9.

$P = 155\ MW \cdot \dfrac{10^6\ W}{1\ MW} = 1.55 \times 10^8\ W$

$V = 762\ V$

$I = ?$

$P = VI$

$I = \dfrac{P}{V} = \dfrac{1.55 \times 10^8\ W}{762\ V} = 203{,}000\ A$

Equivalent Resistance Calculations

1.

$R_4 + R_5 = 0.99\ k + 4.7\ k = 5.69\ k$

$R_2 + R_3 = 1\ k + 3.3\ k = 4.3\ k$

$5.69\ k \parallel 4.3\ k = \dfrac{5.69\ k \cdot 4.3\ k}{5.69\ k + 4.3\ k} = 2.4492\ k$

$R_{EQ} = R_1 + 2.4492\ k = 1\ k + 2.4492\ k = 3.4492\ k\Omega$

2.

$R_2 \parallel R_3 = \dfrac{550\ k \cdot 470\ k}{550\ k + 470\ k} = 253.4314\ k$

$R_{EQ} = R_1 + 253.4314\ k + R_4 = 5.5\ M + 0.2534\ M + 1.2\ M = 6.9534\ M\Omega$

3.

$R_6 \parallel R_7 = 80 \parallel 90 = \dfrac{80 \cdot 90}{80 + 90} = 42.3529$

$R_5 + 42.3529 = 80 + 42.3529 = 122.3529$

$R_3 + R_4 = 40 + 40 = 80$

$80 \parallel 122.3529 = \dfrac{80 \cdot 122.3529}{80 + 122.3529} = 48.3721$

$R_{EQ} = R_1 + R_2 + 48.3721 = 50 + 25 + 48.3721 = 123.3721\ \Omega$

4.

$$R_7 + R_8 = 50 + 50 = 100$$

$$R_6 \| 100 = 50 \| 100 = \frac{50 \cdot 100}{50 + 100} = 33.3333$$

$$R_4 + 33.3333 + R_5 = 10 + 33.3333 + 50 = 93.3333$$

$$R_3 \| 93.3333 = 10 \| 93.3333 = \frac{10 \cdot 93.3333}{10 + 93.3333} = 9.0323$$

$$R_2 \| 9.0323 = 10 \| 9.0323 = \frac{10 \cdot 9.0323}{10 + 9.0323} = 4.7458$$

$$R_{EQ} = R_1 \| 4.7458 = 10 \| 4.7458 = \frac{10 \cdot 4.7458}{10 + 4.7458} = 3.2184 \ \Omega$$

Multi-Resistor Calculations 1

1.

$$R_{EQ} = 1 \text{ k} + 2 \text{ k} = 3 \text{ k}$$

$$I = \frac{V_B}{R_{EQ}} = \frac{6 \text{ V}}{3 \text{ k}} = 2.0000 \text{ mA}$$

$$V_1 = IR_1 = 2.0000 \text{ mA} \cdot 1 \text{ k} = 2.0000 \text{ V}$$

2.

$$I = \frac{V_B}{R_2} = \frac{9 \text{ V}}{8 \text{ k}} = 1.1250 \text{ mA}$$

3.

$$R_2 \| R_3 = \frac{2 \text{ k} \cdot 2 \text{ k}}{2 \text{ k} + 2 \text{ k}} = 1 \text{ k}$$

$$R_{EQ} = R_1 + 1 \text{ k} + R_4 = 2 \text{ k} + 1 \text{ k} + 4 \text{ k} = 7 \text{ k}$$

$$I_1 = \frac{V_B}{R_{EQ}} = \frac{6 \text{ V}}{7 \text{ k}} = 0.8571 \text{ mA}$$

$$V_4 = I_1 R_4 = 0.8571 \text{ mA} \cdot 4 \text{ k} = 3.4284 \text{ V}$$

4.

$$R_2 \parallel R_3 = 10 \text{ k} \parallel 5 \text{ k} = \frac{10 \text{ k} \cdot 5 \text{ k}}{10 \text{ k} + 5 \text{ k}} = 3.3333 \text{ k}$$

$$R_{EQ} = R_1 + 3.3333 \text{ k} = 5 \text{ k} + 3.3333 \text{ k} = 8.3333 \text{ k}$$

$$I_1 = \frac{V_B}{R_{EQ}} = \frac{9 \text{ V}}{8.3333 \text{ k}} = 1.0800 \text{ mA}$$

$$V_1 = I_1 R_1 = 1.0800 \text{ mA} \cdot 5 \text{ k} = 5.4000 \text{ V}$$

$$V_3 = 9 \text{ V} - V_1 = 9 \text{ V} - 5.4000 \text{ V} = 3.6000 \text{ V}$$

$$P_{R3} = \frac{V_3^2}{R_3} = \frac{(3.6000 \text{ V})^2}{5 \text{ k}} = 2.5920 \text{ mW}$$

5.

$$R_{EQ} = R_1 + R_2 + R_3 + R_4 = 1 \text{ k} + 2 \text{ k} + 3 \text{ k} + 4 \text{ k} = 10 \text{ k}$$

$$I_1 = \frac{V_B}{R_{EQ}} = \frac{3 \text{ V}}{10 \text{ k}} = 0.3000 \text{ mA}$$

$$V_1 = I_1 R_1 = 0.3000 \text{ mA} \cdot 1 \text{ k}\Omega = 0.3000 \text{ V}$$

$$V_2 = I_1 R_2 = 0.3000 \text{ mA} \cdot 2 \text{ k}\Omega = 0.6000 \text{ V}$$

$$V_3 = I_1 R_3 = 0.3000 \text{ mA} \cdot 3 \text{ k}\Omega = 0.9000 \text{ V}$$

$$V_4 = I_1 R_4 = 0.3000 \text{ mA} \cdot 4 \text{ k}\Omega = 1.2000 \text{ V}$$

$$P_{R1} = V_1 I_1 = 0.3000 \text{ mA} \cdot 0.3000 \text{ V} = 0.0900 \text{ mW}$$

$$P_{R2} = V_2 I_1 = 0.6000 \text{ mA} \cdot 0.3000 \text{ V} = 0.1800 \text{ mW}$$

$$P_{R3} = V_3 I_1 = 0.9000 \text{ mA} \cdot 0.3000 \text{ V} = 0.2700 \text{ mW}$$

$$P_{R4} = V_4 I_1 = 1.2000 \text{ mA} \cdot 0.3000 \text{ V} = 0.3600 \text{ mW}$$

$$0.09000 \text{ mW} + 0.1800 \text{ mW} + 0.2700 \text{ mW} + 0.3600 \text{ mW} = 0.9000 \text{ mW}$$

$$P_B = V_B I_1 = 3 \text{ V} \cdot 0.3000 \text{ mA} = 0.9000 \text{ mW}$$

$$P_{R1} + P_{R2} + P_{R3} + P_{R4} = P_B$$

Multi-Resistor Calculations 2

1.

$$R_2 \| R_3 = \frac{2.2 \text{ k} \cdot 4.5 \text{ k}}{2.2 \text{ k} + 4.5 \text{ k}} = 1.4776 \text{ k}$$

$$R_{EQ} = R_1 + 1.4776 \text{ k} = 1.5 \text{ k} + 1.4776 \text{ k} = 2.9776 \text{ k}\Omega$$

$$I_1 = \frac{V_B}{R_{EQ}} = \frac{5 \text{ V}}{2.9776 \text{ k}} = 1.6792 \text{ mA}$$

$$V_1 = I_1 R_1 = 1.6792 \text{ mA} \cdot 1.5 \text{ k} = 2.5188 \text{ V}$$

$$V_2 = V_B - V_1 = 5 \text{ V} - 2.5188 \text{ V} = 2.4812 \text{ V}$$

$$I_{R2} = \frac{V_2}{R_2} = \frac{2.4812 \text{ V}}{2.2 \text{ k}} = 1.1278 \text{ mA}$$

2.

$$R_2 + R_3 + R_4 = 0.9 \text{ k} + 1 \text{ k} + 2.1 \text{ k} = 4 \text{ k}$$

$$R_{EQ} = 4.3 \text{ k} \| 4 \text{ k} = \frac{4.3 \text{ k} \cdot 4 \text{ k}}{4.3 \text{ k} + 4 \text{ k}} = 2.0723 \text{ k}\Omega$$

$$I_1 = \frac{V_B}{R_{EQ}} = \frac{9 \text{ V}}{2.0723 \text{ k}\Omega} = 4.3430 \text{ mA}$$

$$I_{R1} = \frac{V_B}{R_1} = \frac{9 \text{ V}}{4.3 \text{ k}} = 2.0930 \text{ mA}$$

$$I_{R2} = I_1 - I_{R1} = 4.3430 \text{ mA} - 2.0930 \text{ mA} = 2.2500 \text{ mA}$$

$$V_3 = I_{R2} R_3 = 2.2500 \text{ mA} \cdot 1.0 \text{ k} = 2.2500 \text{ V}$$

3.

$$R_3 + R_4 = 3.3 \text{ k} + 4.7 \text{ k} = 8 \text{ k}$$

$$R_2 \| 8 \text{ k} = \frac{2 \text{ k} \cdot 8 \text{ k}}{2 \text{ k} + 8 \text{ k}} = 1.6 \text{ k}$$

$$R_{EQ} = R_1 + 1.6 \text{ k} = 0.5 \text{ k} + 1.6 \text{ k} = 2.1000 \text{ k}\Omega$$

$$I_1 = \frac{V_B}{R_{EQ}} = \frac{12 \text{ V}}{2.1000 \text{ k}} = 5.7143 \text{ mA}$$

$$V_1 = I_1 R_1 = 5.7143 \text{ mA} \cdot 0.5 \text{ k} = 2.8572 \text{ V}$$

$$V_2 = V_B - V_1 = 12 \text{ V} - 2.8572 \text{ V} = 9.1428 \text{ V}$$

$$I_{R2} = \frac{V_2}{R_2} = \frac{9.1428 \text{ V}}{2 \text{ k}} = 4.5714 \text{ mA}$$

$$I_{R3} = I_1 - I_{R2} = 5.7143 \text{ mA} - 4.5714 \text{ mA} = 1.1429 \text{ mA}$$

$$P_{R4} = I_{R3}^2 R_4 = (1.1429 \text{ mA})^2 \cdot 4.7 \text{ k} = 6.1392 \text{ mWV}$$

4.

$R_3 + R_4 + R_5 = 1.5 \text{ M} + 1.5 \text{ M} + 3.3 \text{ M} = 6.3 \text{ M}$

$R_2 \,\|\, 6.3 \text{ M} = \dfrac{4.7 \text{ M} \cdot 6.3 \text{ M}}{4.7 \text{ M} + 6.3 \text{ M}} = 2.6918 \text{ M}$

$R_{EQ} = R_1 + 2.6918 \text{ M} + R_8 = 1.5 \text{ M} + 2.6918 \text{ M} + 3.3 \text{ M} = 7.4918 \text{ M}$

$V_B = I_1 R_{EQ}$

$I_1 = \dfrac{V_B}{R_{EQ}} = \dfrac{6 \text{ V}}{7.4918 \text{ M}\Omega} = 0.8009 \ \mu\text{A}$

$V_1 = I_1 R_1 = 0.8009 \ \mu\text{A} \cdot 1.5 \text{ M} = 1.2013 \text{ V}$

$V_6 = I_1 R_6 = 0.8009 \ \mu\text{A} \cdot 3.3 \text{ M} = 2.6430 \text{ V}$

$V_B = V_1 + V_2 + V_6$

$V_2 = V_B - V_1 - V_6 = 6 \text{ V} - 1.2013 \text{ V} - 2.6430 \text{ V} = 2.1557 \text{ V}$

$V_2 = I_2 R_2$

$I_{R2} = \dfrac{V_2}{R_2} = \dfrac{2.1557 \text{ V}}{4.7 \text{ M}} = 0.4587 \ \mu\text{A}$

$I_{R1} = I_{R2} + I_{R3}$

$I_{R3} = I_1 - I_{R2} = 0.8009 \ \mu\text{A} - 0.4587 \ \mu\text{A} = 0.3422 \ \mu\text{A}$

$V_3 = I_{R3} R_3 = 0.3422 \ \mu\text{A} \cdot 1.5 \text{ M} = 0.5133 \text{ V}$

$V_4 = I_{R3} R_4 = 0.3422 \ \mu\text{A} \cdot 1.5 \text{ M} = 0.5133 \text{ V}$

$V_5 = I_{R3} R_5 = 0.3422 \ \mu\text{A} \cdot 3.3 \text{ M} = 1.1293 \text{ V}$

Multi-Resistor Calculations 3

1.

$R_3 \,\|\, R_4 = 3 \text{ k} \,\|\, 3 \text{ k} = 1.5 \text{ k}$

$R_{EQ} = R_1 + R_2 + 1.5 \text{ k} = 2 \text{ k} + 2 \text{ k} + 1.5 \text{ k} = 5.5 \text{ k}\Omega$

$I_1 = \dfrac{V_B}{R_{EQ}} = \dfrac{5.5 \text{ V}}{5.5 \text{ k}\Omega} = 1.0000 \text{ mA}$

$V_1 = I_1 R_1 = 1.0000 \text{ mA} \cdot 2 \text{ k} = 2.0000 \text{ V}$

$V_2 = I_1 R_2 = 1.0000 \text{ mA} \cdot 2 \text{ k} = 2.0000 \text{ V}$

$V_3 = V_B - V_1 - V_2 = 5.5 \text{ V} - 2.0000 \text{ V} - 2.0000 \text{ V} = 1.5000 \text{ V}$

$I_{R3} = \dfrac{V_3}{R_3} = \dfrac{1.5 \text{ V}}{3.0 \text{ k}\Omega} = 0.5000 \text{ mA}$

2.

$$R_{EQ} = (R_1 \| R_2) + R_3 = \frac{4.4 \text{ k} \cdot 5.3 \text{ k}}{4.4 \text{ k} + 5.3 \text{ k}} + 6.1 \text{ k} = 8.5041 \text{ k}\Omega$$

$$I_1 = \frac{V_B}{R_{EQ}} = \frac{8 \text{ V}}{8.5041 \text{ k}\Omega} = 0.9407 \text{ mA}$$

$$V_3 = I_{R3} R_3 = 0.9407 \text{ mA} \cdot 6.1 \text{ k}\Omega = 5.7383 \text{ V}$$

$$V_2 = V_B - V_3 = 8 \text{ V} - 5.7383 \text{ V} = 2.2617 \text{ V}$$

$$I_{R2} = \frac{V_2}{R_2} = \frac{2.2617 \text{ V}}{5.3 \text{ k}} = 0.4267 \text{ mA}$$

3.

$$R_2 + R_3 \| R_5 = 0.8 \text{ k} \| 0.51 \text{ k} = \frac{0.8 \text{ k} \cdot 0.51 \text{ k}}{0.8 \text{ k} + 0.51 \text{ k}} = 0.3115 \text{ k}$$

$$R_{EQ} = R_1 + 0.3114 \text{ k} + R_4 = 0.77 \text{ k} + 0.3115 \text{ k} + 0.87 \text{ k} = 1.9515 \text{ k}\Omega$$

$$I_1 = \frac{V_B}{R_{EQ}} = \frac{12 \text{ V}}{1.9514 \text{ k}\Omega} = 6.1491 \text{ mA}$$

$$V_1 = I_1 R_1 = 6.1491 \text{ mA} \cdot 0.77 \text{ k} = 4.7348 \text{ V}$$

$$V_4 = I_1 R_4 = 6.1491 \text{ mA} \cdot 0.87 \text{ k} = 5.3497 \text{ V}$$

$$V_5 = V_B - V_1 - V_4 = 12 \text{ V} - 4.7348 \text{ V} - 5.3497 \text{ V} = 1.9155 \text{ V}$$

$$I_3 = \frac{V_5}{R_5} = \frac{1.9155 \text{ V}}{0.51 \text{ k}\Omega} = 3.7559 \text{ mA}$$

$$I_2 = I_1 - I_3 = 6.1491 \text{ mA} - 3.7559 \text{ mA} = 2.3932 \text{ mA}$$

$$V_2 = I_{R2} R_2 = 2.3932 \text{ mA} \cdot 0.33 \text{ k} = 0.7898 \text{ V}$$

$$P_{R2} = V_2 I_{R2} = 0.7898 \text{ V} \cdot 2.3932 \text{ mA} = 1.8900 \text{ mW}$$

$$V_3 = I_{R3} R_3 = 2.3932 \text{ mA} \cdot 0.47 \text{ k} = 1.1248 \text{ V}$$

$$P_{R3} = V_3 I_{R3} = 1.1248 \text{ V} \cdot 2.3932 \text{ mA} = 2.6919 \text{ mW}$$

4.

$$R_3 + R_4 + R_5 = 2.4 \text{ k} + 5.1 \text{ k} + 4.7 \text{ k} = 12.2 \text{ k}$$

$$R_2 \, || \, 12.2 \text{ k} = \frac{1.3 \text{ k} \cdot 12.2 \text{ k}}{1.3 \text{ k} + 12.2 \text{ k}} = 1.1748 \text{ k}$$

$$R_{EQ} = R_1 \, || \, 1.1748 \text{ k} = \frac{3.3 \text{ k} \cdot 1.1748 \text{ k}}{3.3 \text{ k} + 1.1748 \text{ k}} = 0.8664 \text{ k}\Omega$$

$$I_1 = \frac{V_B}{R_{EQ}} = \frac{4.7 \text{ V}}{0.8664 \text{ k}\Omega} = 5.4247 \text{ mA}$$

$$I_{R1} = \frac{V_B}{R_1} = \frac{4.7 \text{ V}}{3.3 \text{ k}\Omega} = 1.4242 \text{ mA}$$

$$I_{R2} = \frac{V_B}{R_2} = \frac{4.7 \text{ V}}{1.3 \text{ k}\Omega} = 3.6154 \text{ mA}$$

$$I_{R5} = I_1 - I_{R1} - I_{R2} = 5.4247 \text{ mA} - 1.4242 \text{ mA} - 3.6154 \text{ mA} = 0.3851 \text{ mA}$$

$$V_5 = I_{R5} R_5 = 0.3851 \text{ mA} \cdot 4.7 \text{ k} = 1.8100 \text{ V}$$

$$P_{R5} = V_5 I_{R5} = 1.8100 \text{ V} \cdot 0.3851 \text{ mA} = 0.6970 \text{ mW}$$

Chapter 10

Solubility Calculations

1.

$$0.32 \ \frac{\text{g}}{\text{mL}} = \frac{3,500 \text{ g}}{(x) \text{ mL}}$$

$$(x) \text{ mL} \cdot 0.32 \ \frac{\text{g}}{\text{mL}} = 3,500 \text{ g}$$

$$(x) \text{ mL} = \frac{3,500 \text{ g}}{0.32 \ \frac{\text{g}}{\text{mL}}} = 11,000 \text{ mL} \cdot \frac{1 \text{ L}}{1000 \text{ mL}} = 11 \text{ L}$$

2.

$$1.01 \ \frac{\text{g}}{\text{mL}} \cdot \frac{1 \text{ kg}}{1000 \text{ g}} \cdot \frac{1000 \text{ mL}}{1 \text{ L}} \cdot \frac{1000 \text{ L}}{1 \text{ m}^3} = 1,010 \ \frac{\text{kg}}{\text{m}^3}$$

$$1,010 \ \frac{\text{kg}}{\text{m}^3} = \frac{(x) \text{ kg}}{5.00 \text{ m}^3}$$

$$1,010 \text{ kg} \cdot 5.00 = (x) \text{ kg} = 5,050 \text{ kg}$$

3.

$$\frac{1.750 \text{ kg}}{3.66 \text{ L}} \cdot \frac{1 \text{ L}}{1000 \text{ mL}} \cdot \frac{1000 \text{ g}}{1 \text{ kg}} = 0.478 \ \frac{\text{g}}{\text{mL}}$$

4.

$$\frac{3,610 \text{ kg}}{45,550 \text{ L}} \cdot \frac{1000 \text{ L}}{1 \text{ m}^3} = 79.3 \ \frac{\text{kg}}{\text{m}^3}$$

5.

$$79.3 \ \frac{\text{kg}}{\text{m}^3} = \frac{4.68 \times 10^7 \text{ kg}}{(x) \text{ m}^3}$$

$$(x) \ 79.3 \text{ kg} = 1 \text{ m}^3 \cdot 4.68 \times 10^7 \text{ kg}$$

$$(x) = \frac{4.68 \times 10^7 \text{ kg}}{79.3 \text{ kg}} = 590,164 \text{ m}^3 \cdot \frac{1000 \text{ L}}{1 \text{ m}^3} \cdot \frac{1 \text{ ML}}{10^6 \text{ L}} = 5.90 \times 10^2 \text{ ML}$$

Chapter 11

Volume, Mass, and Weight Exercises

1.

$$98.34 \ \frac{kg}{m^3} \cdot \frac{1000 \ g}{1 \ kg} \cdot \left(\frac{1 \ m}{100 \ cm}\right)^3 = 0.09834 \ \frac{g}{cm^3}$$

2.

$$42 \ mL \cdot \frac{1 \ L}{1000 \ mL} \cdot \frac{1 \ gal}{3.786 \ L} = 0.011 \ gal$$

3.

$$F_w = 18.5 \ lb \cdot \frac{4.45 \ N}{1 \ lb} = 82.3 \ N$$

$$m = ?$$

$$F_w = mg$$

$$m = \frac{F_w}{g} = \frac{82.3 \ N}{9.80 \ \dfrac{m}{s^2}} = 8.40 \ kg$$

4.

$$3.6711 \times 10^4 \ \frac{g}{mL} \cdot \frac{1 \ kg}{1000 \ g} \cdot \frac{1000 \ mL}{1 \ L} \cdot \frac{1000 \ L}{1 \ m^3} = 3.6711 \times 10^7 \ \frac{kg}{m^3}$$

5.

$$1.957 \times 10^4 \ in^3 \cdot \left(\frac{2.54 \ cm}{1 \ in}\right)^3 = 320,700 \ cm^3$$

6.

$$455 \ mL \cdot \frac{1 \ L}{1000 \ mL} \cdot \frac{1 \ m^3}{1000 \ L} = 0.000455 \ m^3$$

7.

$m = 46{,}000$ kg

$F_w = ?$

$F_w = mg = 46{,}000 \text{ kg} \cdot 9.80 \ \dfrac{\text{m}}{\text{s}^2} = 450{,}000 \text{ N}$

$4.508 \times 10^5 \text{ N} \cdot \dfrac{1 \text{ lb}}{4.45 \text{ N}} = 1.0 \times 10^5 \text{ lb}$

8.

$32.11 \text{ L} \cdot \dfrac{1000 \text{ cm}^3}{1 \text{ L}} \cdot \left(\dfrac{1 \text{ in}}{2.54 \text{ cm}}\right)^3 = 1{,}959 \text{ in}^3$

9.

$F_w = 14.89 \text{ N} \cdot \dfrac{1 \text{ lb}}{4.45 \text{ N}} = 3.35 \text{ lb}$

$m = ?$

$m = \dfrac{F_w}{g} = \dfrac{14.89 \text{ N}}{9.80 \ \dfrac{\text{m}}{\text{s}^2}} = 1.52 \text{ kg}$

10.

$36.00 \text{ cm}^3 \cdot \left(\dfrac{1 \text{ m}}{100 \text{ cm}}\right)^3 = 3.6 \times 10^{-5} \text{ m}^3$

11.

$9.11 \text{ m}^3 \cdot \left(\dfrac{100 \text{ cm}}{1 \text{ m}}\right)^3 = 9.11 \times 10^6 \text{ cm}^3$

12.

$4.11 \times 10^5 \text{ m}^3 \cdot \dfrac{1000 \text{ L}}{1 \text{ m}^3} = 4.11 \times 10^8 \text{ L}$

13.

$F_w = 55{,}789 \text{ lb} \cdot \dfrac{4.45 \text{ N}}{1 \text{ lb}} = 2.48261 \times 10^5 \text{ N}$

$m = \dfrac{F_w}{g} = \dfrac{2.48261 \times 10^5 \text{ N}}{9.80 \ \dfrac{\text{m}}{\text{s}^2}} = 25{,}300 \text{ kg}$

14.

$$5.022 \ \frac{g}{cm^3} \cdot \frac{1 \ kg}{1000 \ g} \cdot \left(\frac{100 \ cm}{1 \ m} \right)^3 = 5,022 \ \frac{kg}{m^3}$$

15.

$$3.76 \times 10^{-4} \ \frac{g}{mL} \cdot \frac{1000 \ mL}{1 \ L} \cdot \frac{1 \ L}{1000 \ cm^3} = 3.76 \times 10^{-4} \ \frac{g}{cm^3}$$

16.

$$F_w = 50,000 \ N \cdot \frac{1 \ lb}{4.45 \ N} = 10,000 \ lb$$

$$m = \frac{F_w}{g} = \frac{50,000 \ N}{9.80 \ \frac{m}{s^2}} = 5,000 \ kg$$

17.

$$1.75 \times 10^{-6} \ m^3 \cdot \left(\frac{100 \ cm}{1 \ m} \right)^3 = 1.75 \ cm^3$$

18.

$$100.5 \ ft^3 \cdot \left(\frac{12 \ in}{1 \ ft} \right)^3 \cdot \left(\frac{2.54 \ cm}{1 \ in} \right)^3 \cdot \left(\frac{1 \ m}{100 \ cm} \right)^3 = 2.846 \ m^3$$

19.

$$37 \ m^3 \cdot \left(\frac{100 \ cm}{1 \ m} \right)^3 \cdot \left(\frac{1 \ in}{2.54 \ cm} \right)^3 = 2,300,000 \ in^3$$

20.

$$750 \ cm^3 \cdot \frac{1 \ L}{1000 \ cm^3} = 0.75 \ L$$

21.

$$5,755,000 \ gal \cdot \frac{3.786 \ L}{1 \ gal} \cdot \frac{1000 \ cm^3}{1 \ L} \cdot \left(\frac{1 \ m}{100 \ cm} \right)^3 = 21,790 \ m^3$$

Density Exercises

1.

$m = 0.196$ g

$V = 100.1$ mL

$\rho = ?$

$$\rho = \frac{m}{V} = \frac{0.196 \text{ g}}{100.1 \text{ mL}} = 1.96 \times 10^{-3} \frac{\text{g}}{\text{mL}}$$

2.

$$\rho = 955 \frac{\text{kg}}{\text{m}^3} \cdot \frac{1000 \text{ g}}{1 \text{ kg}} \cdot \left(\frac{1 \text{ m}}{100 \text{ cm}}\right)^3 \cdot \frac{100 \text{ cm}^3}{1 \text{ L}} \cdot \frac{1 \text{ L}}{1000 \text{ mL}} = 0.955 \frac{\text{g}}{\text{mL}}$$

$m = 550$ g

$V = ?$

$$\rho = \frac{m}{V}$$

$$V = \frac{m}{\rho} = \frac{550 \text{ g}}{0.955 \frac{\text{g}}{\text{mL}}} = 580 \text{ mL}$$

3.

$$m = 15.7 \text{ kg} \cdot \frac{1000 \text{ g}}{1 \text{ kg}} = 1.57 \times 10^4 \text{ g}$$

$$\rho = 5.32 \frac{\text{g}}{\text{cm}^3}$$

$V = ?$

$$\rho = \frac{m}{V}$$

$$V = \frac{m}{\rho} = \frac{1.57 \times 10^4 \text{ g}}{5.32 \frac{\text{g}}{\text{cm}^3}} = 2,950 \text{ cm}^3$$

$$2.95 \times 10^3 \text{ cm}^3 \cdot \left(\frac{1 \text{ m}}{100 \text{ cm}}\right)^3 = 0.00295 \text{ m}^3$$

4.

$l = 3.00$ cm

$w = 3.00$ cm

$h = 3.00$ cm

$F_w = 5.336 \times 10^{-2}$ lb $\cdot \dfrac{4.45 \text{ N}}{1 \text{ lb}} = 2.375 \times 10^{-1}$ N

$\rho = ?$

$F_w = mg$

$m = \dfrac{F_w}{g} = \dfrac{2.375 \times 10^{-1} \text{ N}}{9.80 \ \dfrac{\text{m}}{\text{s}^2}} = 2.423 \times 10^{-2} \text{ kg} \cdot \dfrac{1000 \text{ g}}{1 \text{ kg}} = 24.23 \text{ g}$

$V = l \cdot w \cdot h = (3.00 \text{ cm})^3 = 27 \text{ cm}^3$

$\rho = \dfrac{m}{V} = \dfrac{24.23 \text{ g}}{27 \text{ cm}^3} = 0.897 \ \dfrac{\text{g}}{\text{cm}^3}$

5.

$V_1 = 23.35$ mL

$V_2 = 27.79$ mL

$m = 32.1$ g

$\rho = ?$

$V_{Total} = 27.79 \text{ mL} - 23.35 \text{ mL} = 4.44 \text{ mL} \cdot \dfrac{1 \text{ mL}}{1 \text{ cm}^3} = 4.44 \text{ cm}^3$

$\rho = \dfrac{32.1 \text{ g}}{4.44 \text{ cm}^3} = 7.23 \ \dfrac{\text{g}}{\text{cm}^3}$

6.

$$h = 34.5 \text{ in} \cdot \frac{1 \text{ ft}}{12 \text{ in}} \cdot \frac{0.3048 \text{ m}}{1 \text{ ft}} = 0.8763 \text{ m}$$

$$d = 24 \text{ in} \cdot \frac{1 \text{ ft}}{12 \text{ in}} \cdot \frac{0.3048 \text{ m}}{1 \text{ ft}} = 0.6096 \text{ m}$$

$$\rho = 810 \ \frac{\text{kg}}{\text{m}^3}$$

$$m = ?$$

$$V = \pi r^2 h = \pi \cdot \left(\frac{0.6096 \text{ m}}{2} \right)^2 \cdot 0.2558 \text{ m}^3$$

$$\rho = \frac{m}{V}$$

$$m = V\rho = 0.2558 \text{ m}^3 \cdot 810 \ \frac{\text{kg}}{\text{m}^3} = 210 \text{ kg}$$

9.

$$\rho = 7{,}830 \ \frac{\text{kg}}{\text{m}^3} \cdot \frac{1000 \text{ g}}{1 \text{ kg}} \cdot \left(\frac{1 \text{ m}}{100 \text{ cm}} \right)^3 = 7.83 \ \frac{\text{g}}{\text{cm}^3}$$

$$l = 2.1 \text{ cm}$$

$$w = 3.5 \text{ cm}$$

$$m = 94.5 \text{ g}$$

$$h = ?$$

$$\rho = \frac{m}{V}$$

$$V = \frac{m}{\rho} = \frac{94.5 \text{ g}}{7.83 \ \frac{\text{g}}{\text{cm}^3}} = 12.07 \text{ cm}^3$$

$$V = l \cdot w \cdot h$$

$$h = \frac{V}{l \cdot w} = \frac{12.07 \text{ cm}^3}{2.1 \text{ cm} \cdot 3.5 \text{ cm}} = 1.6 \text{ cm}$$

10.

$$m = 306 \text{ g} \cdot \frac{1 \text{ kg}}{1000 \text{ g}} = 0.306 \text{ kg}$$

$$V = 22.5 \text{ mL} \cdot \frac{1 \text{ cm}^3}{1 \text{ mL}} \cdot \left(\frac{1 \text{ m}}{100 \text{ cm}}\right)^3 = 2.25 \times 10^{-5} \text{ m}^3$$

$$\rho = ?$$

$$\rho = \frac{m}{V} = \frac{0.306 \text{ kg}}{2.25 \times 10^{-5} \text{ m}^3} = 13,600 \ \frac{\text{kg}}{\text{m}^3}$$

11.

$$\rho = 10.501 \ \frac{\text{g}}{\text{cm}^3}$$

$$h = 4.500 \text{ cm}$$

$$d = 2.7500 \text{ cm}$$

$$F_w = ?$$

$$V = \pi r^2 h = \pi \cdot \left(\frac{2.7500 \text{ cm}}{2}\right)^2 \cdot 4.500 \text{ cm} = 26.728 \text{ cm}^3$$

$$\rho = \frac{m}{V}$$

$$m = V\rho = 26.728 \text{ cm}^3 \cdot 10.501 \ \frac{\text{g}}{\text{cm}^3} = 280.67 \text{ g} \cdot \frac{1 \text{ kg}}{1000 \text{ g}} = 0.28067$$

$$F_w = mg = 0.28067 \cdot 9.80 \ \frac{\text{m}}{\text{s}^2} = 2.7506 \text{ N} \cdot \frac{1 \text{ lb}}{4.45 \text{ N}} = 0.618 \text{ lb}$$

12.

$$\rho = 1,000.0 \ \frac{\text{kg}}{\text{m}^3}$$

$$V = 5.6 \text{ L} \cdot \frac{1000 \text{ cm}^3}{1 \text{ L}} \cdot \left(\frac{1 \text{ m}}{100 \text{ cm}}\right)^3 = 5.6 \times 10^{-3} \text{ m}^3$$

$$m = ?$$

$$\rho = \frac{m}{V}$$

$$m = V\rho = 5.6 \times 10^{-3} \text{ m}^3 \cdot 1,000.0 \ \frac{\text{kg}}{\text{m}^3} = 5.6 \text{ kg}$$

13.

$$V = 3.0\times10^6 \text{ gal} \cdot \frac{3.786 \text{ L}}{1 \text{ gal}} \cdot \frac{1 \text{ m}^3}{1000 \text{ L}} = 1.14\times10^4 \text{ m}^3$$

$$\rho = 998 \ \frac{\text{kg}}{\text{m}^3}$$

$$F_w = ?$$

$$\rho = \frac{m}{V}$$

$$m = V\rho = 1.14\times10^4 \text{ m}^3 \cdot 998 \ \frac{\text{kg}}{\text{m}^3} = 1.13\times10^7 \text{ kg}$$

$$F_w = mg = 1.13\times10^7 \text{ kg} \cdot 9.80 \ \frac{\text{m}}{\text{s}^2} = 1.11\times10^8 \text{ N}$$

$$F_w = 1.11\times10^8 \text{ N} \cdot \frac{1 \text{ lb}}{4.45 \text{ N}} = 25,000,000 \text{ lb}$$

15.

$$\rho = 7.85 \ \frac{\text{g}}{\text{cm}^3} \cdot \frac{1 \text{ kg}}{1000 \text{ g}} \cdot \left(\frac{100 \text{ cm}}{1 \text{ m}}\right)^3 = 7,850 \ \frac{\text{kg}}{\text{m}^3}$$

$$d = 6.1875 \text{ in} \cdot \frac{1 \text{ ft}}{12 \text{ in}} \cdot \frac{0.3048 \text{ m}}{1 \text{ ft}} = 0.15716 \text{ m}$$

$$h = 7.0000 \text{ in} \cdot \frac{1 \text{ ft}}{12 \text{ in}} \cdot \frac{0.3048 \text{ m}}{1 \text{ ft}} = 0.17780 \text{ m}$$

$$F_w = ?$$

$$V = \pi r^2 h = \pi \cdot \left(\frac{0.15716 \text{ m}}{2}\right)^2 \cdot 0.1778 \text{ m} = 0.0034491 \text{ m}^3$$

$$\rho = \frac{m}{V}$$

$$m = V\rho = 0.0034491 \text{ m}^3 \cdot 7,850 \ \frac{\text{kg}}{\text{m}^3} = 27.075 \text{ kg}$$

$$F_w = mg = 27.075 \text{ kg} \cdot 9.80 \ \frac{\text{m}}{\text{s}^2} = 265.34 \text{ N} \cdot \frac{1 \text{ lb}}{4.45 \text{ N}} = 59.6 \text{ lb}$$

16.

$$\rho = 160 \ \frac{kg}{m^3}$$

$$d = 2.0 \ ft \cdot \frac{0.3048 \ m}{1 \ ft} = 0.6096 \ m$$

$$h = 45 \ ft \cdot \frac{0.3048 \ m}{1 \ ft} = 13.716 \ m$$

$$V = \pi r^2 h = \pi \cdot \left(\frac{0.6096 \ m}{2} \right)^2 \cdot 13.716 \ m = 4.0032 \ m^3$$

$$\rho = \frac{m}{V}$$

$$m = V\rho = 4.0032 \ m^3 \cdot 160 \ \frac{kg}{m^3} = 640.51 \ kg$$

$$F_w = mg = 640.51 \ kg \cdot 9.80 \ \frac{m}{s^2} = 6,277.0 \ N$$

$$9 \ trees \cdot 6,277.0 \ \frac{N}{tree} \cdot \frac{1 \ lb}{4.45 \ N} = 13,000 \ lb$$

17.

$$l = 50.0 \ m$$

$$w = 25.0 \ m$$

$$d = 2.00 \ m$$

$$\rho = 998 \ \frac{kg}{m^3}$$

$$V = ?$$

$$F_w = ?$$

$$V = lwd = 50.0 \ m \cdot 25.0 \ m \cdot 2.00 \ m = 2,500.0 \ m^3 \cdot \frac{1000 \ L}{1 \ m^3} \cdot \frac{1 \ gal}{3.786 \ L} = 6.60 \times 10^5 \ gal$$

$$\rho = \frac{m}{V}$$

$$m = V\rho = 2,500.0 \ m^3 \cdot 998 \ \frac{kg}{m^3} = 2.495 \times 10^6 \ kg$$

$$F_w = mg = 2.495 \times 10^6 \ kg \cdot 9.80 \ \frac{m}{s^2} = 2.4451 \times 10^7 \ N \cdot \frac{1 \ lb}{4.45 \ N} \cdot \frac{1 \ ton}{2,000 \ lb} = 2,750 \ tons$$

Chapter 13

Atoms and Atomic Bonding

3.

$5 \cdot 1.673 \times 10^{-27} \ \text{kg} = 8.365 \times 10^{-27} \ \text{kg}$

$5 \cdot 9.11 \times 10^{-31} \ \text{kg} = 4.555 \times 10^{-30} \ \text{kg}$

$6 \cdot 1.675 \times 10^{-27} \ \text{kg} = 1.005 \times 10^{-26} \ \text{kg}$

$8.365 \times 10^{-27} \ \text{kg} + 4.555 \times 10^{-30} \ \text{kg} + 1.005 \times 10^{-26} \ \text{kg} = 1.8419555 \times 10^{-26} \ \text{kg}$

www.ingramcontent.com/pod-product-compliance
Lightning Source LLC
Chambersburg PA
CBHW081550220326
41598CB00036B/6628

Preface

This solutions manual contains fully detailed solutions for all of the computational problems contained in my text *Principles of Chemistry*. Teachers and students using that text should find this manual to be a valuable resource.

Explanations for the number of significant digits shown in results are included for representative problems. Throughout this manual, significant digits are referred to as "sig digs."

When comparing your results to the results shown here and to those in the text, keep in mind that the last digit is always uncertain because of the way significant digits in measurements are defined. When two results match except for a difference of 1 in the most precise digit, we say that the results match. Because of rounding in calculators, it will not be uncommon for results shown here to differ from the answer key in the text or from your result by 1 in the most precise digit.

I have checked and double checked the solutions to make them as accurate as possible. However, in any manual of this kind it is inevitable that errors remain. If you find an error, we would be much obliged if you would inform us of it by sending an email to info@novarescienceandmath.com.

Chapter 1

8. a.

$$12.55 \text{ ft} \cdot \frac{1 \text{ yd}}{3 \text{ ft}} = 4.183 \text{ yd}$$

8. b.

$$0.44556 \text{ mi} \cdot \frac{5280 \text{ ft}}{1 \text{ mi}} = 2352.6 \text{ ft}$$

8. c.

$$147.55 \text{ in} \cdot \frac{1 \text{ ft}}{12 \text{ in}} = 12.296 \text{ ft}$$

8. d.

$$55.08 \text{ gal} \cdot \frac{1 \text{ ft}^3}{7.4805 \text{ gal}} = 7.363 \text{ ft}^3$$

8. e.

$$934 \text{ ft}^3 \cdot \frac{12 \text{ in}}{1 \text{ ft}} \cdot \frac{12 \text{ in}}{1 \text{ ft}} \cdot \frac{12 \text{ in}}{1 \text{ ft}} = 1,610,000 \text{ in}^3$$

The result is rounded to 3 sig digs because the given information has only 3 sig digs.

8. f.

$$739.22 \frac{\text{ft}^3}{\text{s}} \cdot \frac{7.48052 \text{ gal}}{1 \text{ ft}^3} \cdot \frac{3600 \text{ s}}{1 \text{ hr}} = 19,907,000 \frac{\text{gal}}{\text{hr}}$$

The result is rounded to 5 sig digs because the given information has 5 sig digs.

8. g.

$$12.4 \text{ yr} \cdot \frac{365 \text{ dy}}{1 \text{ yr}} \cdot \frac{24 \text{ hr}}{1 \text{ dy}} = 109,000 \text{ hr}$$

The result is rounded to 3 sig digs because the given information has only 3 sig digs.

8. h.

$$51,083 \text{ in} \cdot \frac{1 \text{ ft}}{12 \text{ in}} \cdot \frac{1 \text{ mi}}{5280 \text{ ft}} = 0.80623 \text{ mi}$$

8. i.

$$14,560.77 \frac{\text{gal}}{\text{hr}} \cdot \frac{4 \text{ qt}}{1 \text{ gal}} \cdot \frac{1 \text{ hr}}{3600 \text{ s}} = 16.17863 \frac{\text{qt}}{\text{s}}$$

8. j.

$$15.90 \ \frac{\text{mi}}{\text{dy}} \cdot \frac{5280 \text{ ft}}{1 \text{ mi}} \cdot \frac{12 \text{ in}}{1 \text{ ft}} \cdot \frac{1 \text{ dy}}{24 \text{ hr}} = 41,980 \ \frac{\text{in}}{\text{hr}}$$

---

9. a.

$$35.4 \text{ mm} \cdot \frac{1 \text{ m}}{1000 \text{ mm}} = 0.0354 \text{ m}$$

---

9. b.

$$76.991 \text{ mL} \cdot \frac{1 \text{ L}}{1000 \text{ mL}} \cdot \frac{10^6 \ \mu\text{L}}{1 \text{ L}} = 76,991 \ \mu\text{L}$$

---

9. c.

$$34.44 \text{ cm}^3 \cdot \frac{1 \text{ mL}}{1 \text{ cm}^3} \cdot \frac{1 \text{ L}}{1000 \text{ mL}} = 0.03444 \text{ L}$$

---

9. d.

$$6.33 \ \frac{\text{g}}{\text{cm}^2} \cdot \frac{1 \text{ kg}}{1000 \text{ g}} \cdot \frac{100 \text{ cm}}{1 \text{ m}} \cdot \frac{100 \text{ cm}}{1 \text{ m}} = 63.3 \ \frac{\text{kg}}{\text{m}^2}$$

---

9. e.

$$9.35 \ \frac{\text{m}}{\text{s}^2} \cdot \frac{1000 \text{ mm}}{1 \text{ m}} \cdot \frac{1 \text{ s}}{1000 \text{ ms}} \cdot \frac{1 \text{ s}}{1000 \text{ ms}} = 0.00935 \ \frac{\text{mm}}{\text{ms}^2}$$

---

9. f.

$$542.2 \ \frac{\text{mJ}}{\text{s}} \cdot \frac{1 \text{ J}}{1000 \text{ mJ}} = 0.5422 \ \frac{\text{J}}{\text{s}}$$

---

9. g.

$$56.6 \ \mu\text{s} \cdot \frac{1 \text{ s}}{10^6 \ \mu\text{s}} \cdot \frac{10^3 \text{ ms}}{1 \text{ s}} = 0.0566 \text{ ms}$$

---

9. h.

$$44.19 \text{ mL} \cdot \frac{1 \text{ cm}^3}{1 \text{ mL}} = 44.19 \text{ cm}^3$$

---

9. i.

$$532 \text{ nm} \cdot \frac{1 \text{ m}}{10^9 \text{ nm}} \cdot \frac{10^6 \ \mu\text{m}}{1 \text{ m}} = 0.532 \ \mu\text{m}$$

---

9. j.

$$96,963,000 \ \frac{\text{mL}}{\text{ms}} \cdot \frac{1 \text{ L}}{1000 \text{ mL}} \cdot \frac{1 \text{ m}^3}{1000 \text{ L}} \cdot \frac{1000 \text{ ms}}{1 \text{ s}} = 96,963 \ \frac{\text{m}^3}{\text{s}}$$

---

9. k.

$$295.6 \text{ cL} \cdot \frac{1 \text{ L}}{100 \text{ cL}} \cdot \frac{10^6 \text{ } \mu\text{L}}{1 \text{ L}} = 2{,}956{,}000 \text{ } \mu\text{L}$$

9. l.

$$0.007873 \text{ m}^3 \cdot \frac{100 \text{ cm}}{1 \text{ m}} \cdot \frac{100 \text{ cm}}{1 \text{ m}} \cdot \frac{100 \text{ cm}}{1 \text{ m}} \cdot \frac{1 \text{ mL}}{1 \text{ cm}^3} = 7873 \text{ mL}$$

9. m.

$$8750 \text{ mm}^2 \cdot \frac{1 \text{ m}}{1000 \text{ mm}} \cdot \frac{1 \text{ m}}{1000 \text{ mm}} = 0.00875 \text{ m}^2$$

9. n.

$$87.1 \frac{\text{cm}}{\text{s}^2} \cdot \frac{1 \text{ m}}{100 \text{ cm}} = 0.871 \frac{\text{m}}{\text{s}^2}$$

9. o.

$$15.75 \frac{\text{kg}}{\text{m}^3} \cdot \frac{1000 \text{ g}}{1 \text{ kg}} \cdot \frac{1 \text{ m}}{100 \text{ cm}} \cdot \frac{1 \text{ m}}{100 \text{ cm}} \cdot \frac{1 \text{ m}}{100 \text{ cm}} = 0.01575 \frac{\text{g}}{\text{cm}^3}$$

9. p.

$$0.875 \text{ km} \cdot \frac{1000 \text{ m}}{1 \text{ km}} = 875 \text{ m}$$

9. q.

$$16{,}056 \text{ MPa} \cdot \frac{10^6 \text{ Pa}}{1 \text{ MPa}} \cdot \frac{1 \text{ kPa}}{10^3 \text{ Pa}} = 16{,}056{,}000 \text{ kPa}$$

9. r.

$$7845 \text{ } \mu\text{A} \cdot \frac{1 \text{ A}}{10^6 \text{ } \mu\text{A}} \cdot \frac{1000 \text{ mA}}{1 \text{ A}} = 7.845 \text{ mA}$$

10. a.

$$T_C = \frac{5}{9}\left(T_F - 32\right) = \frac{5}{9}\left(431.1 - 32\right) = 221.7°\text{C}$$

The 32 is exact, so it does not limit the sig digs resulting from the subtraction. The 5/9 is also exact, so the result has 4 sig digs, just like the given value.

$$T_K = T_C + 273.15 = 221.7 + 273.15 = 494.9 \text{ K}$$

By the addition rule, the result is limited to one decimal.

10. b.

$$T_F = \frac{9}{5}T_C + 32 = \frac{9}{5}(-56.1) + 32 = -69.0°F$$

$$T_K = T_C + 273.15 = -56.1 + 273.15 = 217.1 \text{ K}$$

10. c.

$$T_K = T_C + 273.15 \rightarrow T_C = T_K - 273.15 = 16.0 - 273.15 = -257.2°C$$

The given value limits the result to one decimal place because of the addition rule.

$$T_F = \frac{9}{5}T_C + 32 = \frac{9}{5}(-257.2) + 32 = -431.0°F$$

10. d.

$$T_C = \frac{5}{9}(T_F - 32) = \frac{5}{9}(0.0 - 32) = -17.7\overline{7}°C$$

Since the given value is exact and the conversion is exact, the result is also exact, indicated by the repeating decimal (the bar over the 7). This result can be written with as many decimals as you please.

$$T_K = T_C + 273.15 = -17.7\overline{7} + 273.15 = 255.37\overline{2} \text{ K (exact)}$$

Since both values are exact, the addition rule places no limit on the precision indicated in the result. The only way to indicate that this result is exact is to say so. No amount of zeros added to the end of the value will indicate that the value is exact.

10. e.

$$T_K = T_C + 273.15 = -77.0 + 273.15 = 196.2 \text{ K}$$

$$T_F = \frac{9}{5}T_C + 32 = \frac{9}{5}(-77.0) + 32 = -107°F$$

The result is limited to 3 sig digs because the given value has 3 sig digs.

10. f.

$$T_K = T_C + 273.15 \rightarrow T_C = T_K - 273.15 = 4002 - 273.15 = 3729°C$$

$$T_F = \frac{9}{5}T_C + 32 = \frac{9}{5}(3729) + 32 = 6744°F$$

10. g.

$$T_C = \frac{5}{9}(T_F - 32) = \frac{5}{9}(-32.0 - 32) = -35.6°C$$

$$T_K = T_C + 273.15 = -35.6 + 273.15 = 237.6 \text{ K}$$

10. h.

$$T_K = T_C + 273.15 = 65.25 + 273.15 = 338.40 \text{ K}$$

$$T_F = \frac{9}{5}T_C + 32 = \frac{9}{5}(65.25) + 32 = 149.5°\text{F}$$

10. i.

$$T_K = T_C + 273.15 = 1958 + 273.15 = 2231 \text{ K}$$

$$T_F = \frac{9}{5}T_C + 32 = \frac{9}{5}(1958) + 32 = 3556°\text{F}$$

10. j.

$$T_K = T_C + 273.15 \rightarrow T_C = T_K - 273.15 = 998.0 - 273.15 = 724.8°\text{C}$$

$$T_F = \frac{9}{5}T_C + 32 = \frac{9}{5}(724.8) + 32 = 1337°\text{F}$$

18. a.

$$\frac{|0.239 \text{ g} - 0.234 \text{ g}|}{0.239 \text{ g}} \times 100\% = \frac{0.005}{0.239} \times 100\% = 2\%$$

$$\frac{|1.688 \text{ g} - 1.678 \text{ g}|}{1.688 \text{ g}} \times 100\% = \frac{0.010}{1.688} \times 100\% = 0.59\%$$

$$\frac{|4.678 \text{ g} - 4.446 \text{ g}|}{4.678 \text{ g}} \times 100\% = \frac{0.232}{4.678} \times 100\% = 4.96\%$$

18. b.

$$\frac{\left|2.72 \frac{\text{g}}{\text{cm}^3} - 2.81 \frac{\text{g}}{\text{cm}^3}\right|}{2.72 \frac{\text{g}}{\text{cm}^3}} \times 100\% = \frac{0.09}{2.72} \times 100\% = 3\%$$

18. c

$$\frac{\left|12.011 \frac{\text{g}}{\text{mol}} - 12.0117 \frac{\text{g}}{\text{mol}}\right|}{12.011 \frac{\text{g}}{\text{mol}}} \times 100\% = \frac{0.001}{12.011} \times 100\% = 0.008\%$$

18. d.

$$\frac{|23.4 \text{ kg} - 21.610 \text{ kg}|}{23.4 \text{ kg}} \times 100\% = \frac{1.8}{23.4} \times 100\% = 7.7\%$$

$$\frac{|2.21 \text{ kg} - 1.995 \text{ kg}|}{2.21 \text{ kg}} \times 100\% = \frac{0.22}{2.21} \times 100\% = 10\%$$

To express the second result with the required 2 sig digs, we must write it as $1.0 \times 10^1\%$.

19. a.

$$1737 \text{ km} \cdot \frac{1000 \text{ m}}{1 \text{ km}} \cdot \frac{1 \text{ ft}}{0.3048 \text{ m}} = 5,699,000 \text{ ft} = 5.699 \times 10^6 \text{ ft}$$

$$(\text{Or: } 1737 \text{ km} \cdot \frac{1000 \text{ m}}{1 \text{ km}} \cdot \frac{1 \text{ mi}}{1609 \text{ m}} \cdot \frac{5280 \text{ ft}}{1 \text{ mi}} = 5.700 \times 10^6 \text{ ft})$$

19. b.

$$2.20 \text{ g} \cdot \frac{1 \text{ kg}}{1000 \text{ g}} = 0.00220 \text{ kg} = 2.20 \times 10^{-3} \text{ kg}$$

19. c.

$$591 \text{ mL} \cdot \frac{1 \text{ L}}{1000 \text{ mL}} \cdot \frac{10^6 \text{ μL}}{1 \text{ L}} = 591,000 \text{ μL} = 5.91 \times 10^5 \text{ μL}$$

19. d.

$$7 \times 10^8 \text{ m} \cdot \frac{1 \text{ mi}}{1609 \text{ m}} = 400,000 \text{ mi} = 4 \times 10^5 \text{ mi}$$

19. e.

$$1.616 \times 10^{-35} \text{ m} \cdot \frac{1 \text{ ft}}{0.3048 \text{ m}} = 5.302 \times 10^{-35} \text{ ft}$$

19. f.

$$750 \text{ cm}^3 \cdot \frac{1 \text{ m}}{100 \text{ cm}} \cdot \frac{1 \text{ m}}{100 \text{ cm}} \cdot \frac{1 \text{ m}}{100 \text{ cm}} = 0.00075 \text{ m}^3 = 7.5 \times 10^{-4} \text{ m}^3$$

19. g.

$$2.9979 \times 10^8 \frac{\text{m}}{\text{s}} \cdot \frac{1 \text{ ft}}{0.3048 \text{ m}} \cdot \frac{1 \text{ mi}}{5280 \text{ ft}} \cdot \frac{3600 \text{ s}}{1 \text{ hr}} = 6.7061 \times 10^8 \frac{\text{mi}}{\text{hr}}$$

By using the exact factor 0.3048 m = 1 ft instead of the approximate factor 1609 m = 1 mi we preserve the 5 sig digs in our result.

19. h.

$$168 \text{ hr} \cdot \frac{3600 \text{ s}}{1 \text{ hr}} = 605{,}000 \text{ s} = 6.05 \times 10^5 \text{ s}$$

If the problem had said exactly one week, we would calculate 604,800 s. But we were given a number of hours with 3 sig digs, so that is what we have in the result.

19. i.

$$5570 \ \frac{\text{kg}}{\text{m}^3} \cdot \frac{1000 \text{ g}}{1 \text{ kg}} \cdot \frac{1 \text{ m}}{100 \text{ cm}} \cdot \frac{1 \text{ m}}{100 \text{ cm}} \cdot \frac{1 \text{ m}}{100 \text{ cm}} = 5.57 \ \frac{\text{g}}{\text{cm}^3}$$

19. j.

$$45 \ \frac{\text{gal}}{\text{s}} \cdot \frac{3.785 \text{ L}}{1 \text{ gal}} \cdot \frac{1 \text{ m}^3}{1000 \text{ L}} \cdot \frac{60 \text{ s}}{1 \text{ min}} = 10 \ \frac{\text{m}^3}{\text{min}}$$

To express this result with the required 2 sig digs, we must write it as 1.0×10^1 m³/min.

19. k.

$$600{,}000 \ \frac{\text{ft}^3}{\text{s}} \cdot \frac{12 \text{ in}}{1 \text{ ft}} \cdot \frac{12 \text{ in}}{1 \text{ ft}} \cdot \frac{12 \text{ in}}{1 \text{ ft}} \cdot \frac{2.54 \text{ cm}}{1 \text{ in}} \cdot \frac{2.54 \text{ cm}}{1 \text{ in}} \cdot \frac{2.54 \text{ cm}}{1 \text{ in}} \cdot \frac{1 \text{ L}}{1000 \text{ cm}^3} \cdot \frac{3600 \text{ s}}{1 \text{ hr}} = 6 \times 10^{10} \ \frac{\text{L}}{\text{hr}}$$

19. l.

$$5200 \text{ mL} \cdot \frac{1 \text{ cm}^3}{1 \text{ mL}} \cdot \frac{1 \text{ m}}{100 \text{ cm}} \cdot \frac{1 \text{ m}}{100 \text{ cm}} \cdot \frac{1 \text{ m}}{100 \text{ cm}} = 0.0052 \text{ m}^3 = 5.2 \times 10^{-3} \text{ m}^3$$

19. m.

$$5.65 \times 10^2 \text{ mm}^2 \cdot \frac{1 \text{ m}}{1000 \text{ mm}} \cdot \frac{1 \text{ m}}{1000 \text{ mm}} \cdot \frac{100 \text{ cm}}{1 \text{ m}} \cdot \frac{100 \text{ cm}}{1 \text{ m}} \cdot \frac{1 \text{ in}}{2.54 \text{ cm}} \cdot \frac{1 \text{ in}}{2.54 \text{ cm}} = 0.876 \text{ in}^2$$

19. n.

$$32.16 \ \frac{\text{ft}}{\text{s}^2} \cdot \frac{0.3048 \text{ m}}{1 \text{ ft}} = 9.802 \ \frac{\text{m}}{\text{s}^2}$$

19. o.

$$10.6 \ \mu\text{m} \cdot \frac{1 \text{ m}}{10^6 \ \mu\text{m}} \cdot \frac{100 \text{ cm}}{1 \text{ m}} \cdot \frac{1 \text{ in}}{2.54 \text{ cm}} = 0.000417 \text{ in} = 4.17 \times 10^{-4} \text{ in}$$

19. p.

$$1.1056 \ \frac{\text{g}}{\text{mL}} \cdot \frac{1 \text{ kg}}{10^3 \text{ g}} \cdot \frac{1 \text{ mL}}{1 \text{ cm}} \cdot \frac{100 \text{ cm}}{1 \text{ m}} \cdot \frac{100 \text{ cm}}{1 \text{ m}} \cdot \frac{100 \text{ cm}}{1 \text{ m}} = 1105.6 \ \frac{\text{kg}}{\text{m}^3} = 1.1056 \times 10^3 \ \frac{\text{kg}}{\text{m}^3}$$

19. q.

$$13.6 \ \frac{\text{g}}{\text{cm}^3} \cdot \frac{1000 \text{ mg}}{1 \text{ g}} \cdot \frac{100 \text{ cm}}{1 \text{ m}} \cdot \frac{100 \text{ cm}}{1 \text{ m}} \cdot \frac{100 \text{ cm}}{1 \text{ m}} = 1.36 \times 10^{10} \ \frac{\text{mg}}{\text{m}^3}$$

19. r.

$$93{,}000{,}000 \text{ mi} \cdot \frac{1609 \text{ m}}{1 \text{ mi}} \cdot \frac{100 \text{ cm}}{1 \text{ m}} = 1.5 \times 10^{13} \text{ cm}$$

19. s.

$$65 \frac{\text{mi}}{\text{hr}} \cdot \frac{1609 \text{ m}}{1 \text{ mi}} \cdot \frac{1 \text{ hr}}{3600 \text{ s}} = 29 \frac{\text{m}}{\text{s}}$$

19. t.

$$633 \text{ nm} \cdot \frac{1 \text{ m}}{10^9 \text{ nm}} \cdot \frac{100 \text{ cm}}{1 \text{ m}} \cdot \frac{1 \text{ in}}{2.54 \text{ cm}} = 0.0000249 \text{ in} = 2.49 \times 10^{-5} \text{ in}$$

19. u.

First convert the speed of light to mi/hr:

$$2.9979 \times 10^8 \frac{\text{m}}{\text{s}} \cdot \frac{1 \text{ mi}}{1609 \text{ m}} \cdot \frac{3600 \text{ s}}{1 \text{ hr}} = 6.708 \times 10^8 \frac{\text{mi}}{\text{hr}}$$

Now calculate the proportion indicated by the percentage:

$$6.708 \times 10^8 \frac{\text{mi}}{\text{hr}} \cdot 0.05015 = 3.364 \times 10^7 \frac{\text{mi}}{\text{hr}} = 33{,}640{,}000 \frac{\text{mi}}{\text{hr}}$$

19. v.

$$6.01 \frac{\text{kJ}}{\text{mol}} \cdot \frac{1000 \text{ J}}{1 \text{ kJ}} = 6010 \frac{\text{J}}{\text{mol}} = 6.01 \times 10^3 \frac{\text{J}}{\text{mol}}$$

19. w.

$$32.1 \text{ bar} \cdot \frac{100{,}000 \text{ Pa}}{1 \text{ bar}} \cdot \frac{1 \text{ psi}}{6895 \text{ Pa}} = 466 \text{ psi} = 4.66 \times 10^2 \text{ psi}$$

19. x.

$$0.116 \text{ nm} \cdot \frac{1 \text{ m}}{10^9 \text{ nm}} \cdot \frac{100 \text{ cm}}{1 \text{ m}} = 1.16 \times 10^{-8} \text{ cm}$$

19. y.

$$6.54 \times 10^{-24} \text{ cm}^3 \cdot \frac{1 \text{ in}}{2.54 \text{ cm}} \cdot \frac{1 \text{ in}}{2.54 \text{ cm}} \cdot \frac{1 \text{ in}}{2.54 \text{ cm}} = 3.99 \times 10^{-25} \text{ in}^3$$

19. z.

$$0.385 \frac{\text{J}}{\text{g} \cdot \text{K}} \cdot \frac{1 \text{ g}}{1000 \text{ mg}} = 0.000385 \frac{\text{J}}{\text{mg} \cdot \text{K}} = 3.85 \times 10^{-4} \frac{\text{J}}{\text{mg} \cdot \text{K}}$$

19. aa.

$$370 \text{ mL} \cdot \frac{1 \text{ cm}^3}{1 \text{ mL}} \cdot \frac{1 \text{ in}}{2.54 \text{ cm}} \cdot \frac{1 \text{ in}}{2.54 \text{ cm}} \cdot \frac{1 \text{ in}}{2.54 \text{ cm}} \cdot \frac{1 \text{ ft}}{12 \text{ in}} \cdot \frac{1 \text{ ft}}{12 \text{ in}} \cdot \frac{1 \text{ ft}}{12 \text{ in}} = 0.013 \text{ ft}^3 = 1.3 \times 10^{-2} \text{ ft}^3$$

19. ab.

$$268,581 \text{ mi}^2 \cdot \frac{5280 \text{ ft}}{1 \text{ mi}} \cdot \frac{5280 \text{ ft}}{1 \text{ mi}} \cdot \frac{0.3048 \text{ m}}{1 \text{ ft}} \cdot \frac{0.3048 \text{ m}}{1 \text{ ft}} \cdot \frac{1000 \text{ mm}}{1 \text{ m}} \cdot \frac{1000 \text{ mm}}{1 \text{ m}} = 6.95622 \times 10^{17} \text{ mm}^2$$

19. ac.

$$50,200 \frac{\text{mi}^2}{\text{yr}} \cdot \frac{5280 \text{ ft}}{1 \text{ mi}} \cdot \frac{5280 \text{ ft}}{1 \text{ mi}} \cdot \frac{1 \text{ yr}}{365 \text{ dy}} \cdot \frac{1 \text{ dy}}{24 \text{ hr}} \cdot \frac{1 \text{ hr}}{3600 \text{ s}} = 44,400 \frac{\text{ft}^2}{\text{s}} = 4.44 \times 10^4 \frac{\text{ft}^2}{\text{s}}$$

Chapter 2

12.

Sig digs in all products are determined by the factor with the lowest precision. Sig digs in all sums are determined according to the addition rule.

silicon

	25.8011 u
27.9769 u·0.92223 = 25.8011 u	1.3585 u
28.9765 u·0.04685 = 1.3585 u	+ 0.92679 u
29.9738 u·0.03092 = 0.92679 u	28.086 u

calcium

	38.740 u
39.9626 u·0.96941 = 38.740 u	+ 0.272 u
41.9856 u·0.00647 = 2.72 u	39.012 u

iron

	3.153 u
53.9396 u·0.05845 = 3.153 u	51.323 u
55.9349 u·0.91754 = 51.323 u	1.206 u
56.9354 u·0.02119 = 1.206 u	+ 0.163 u
57.9333 u·0.00282 = 0.163 u	55.822 u

uranium

	1.693 u
235.0439 u·0.007204 = 1.693 u	+ 236.323 u
238.0508 u·0.992742 = 236.323 u	238.016 u

14.

In Table 2.6, the heaviest nuclide listed is U-238, $Z = 92$, so there are 92 protons.

238 – 92 = 146 neutrons

92 electrons

In Table 2.6, the lightest nuclide listed is H-1, $Z = 1$, so there is 1 proton.

1 – 1 = 0 neutrons

1 electron

15.

Using data from Tables 2.6 and 2.7:

92 protons	$92 \cdot 1.007276 \text{ u} = 92.66939 \text{ u}$
146 neutrons	$146 \cdot 1.008665 \text{ u} = 147.2651 \text{ u}$
92 electrons	$92 \cdot 0.0005486 \text{ u} = 0.05047 \text{ u}$

$$
\begin{array}{r}
92.66939 \text{ u} \\
147.2651 \text{ u} \\
+ \quad 0.05047 \text{ u} \\
\hline
239.9850 \text{ u}
\end{array}
$$

$$
\begin{array}{r}
239.9850 \text{ u} \\
- \; 238.0508 \text{ u} \\
\hline
1.9342 \text{ u}
\end{array}
$$

$$1.9342 \text{ u} = 1.9342 \; \frac{\text{g}}{\text{mol}} \cdot \frac{\text{mol}}{6.02214076 \times 10^{23} \text{ atoms}} = 3.2118 \times 10^{-24} \; \frac{\text{g}}{\text{atom}}$$

16.

$m = 0.196 \text{ g}$

$V = 100.1 \text{ mL}$

$$\rho = \frac{m}{V} = \frac{0.196 \text{ g}}{100.1 \text{ mL}} = 0.00196 \; \frac{\text{g}}{\text{mL}}$$

17.

$m = 550 \text{ g}$

$$\rho = 955 \; \frac{\text{kg}}{\text{m}^3} \cdot \frac{1000 \text{ g}}{1 \text{ kg}} \cdot \frac{1 \text{ m}^3}{1000 \text{ L}} \cdot \frac{1 \text{ L}}{1000 \text{ mL}} = 0.955 \; \frac{\text{g}}{\text{mL}}$$

$$\rho = \frac{m}{V} \rightarrow V = \frac{m}{\rho} = \frac{550 \text{ g}}{0.955 \; \frac{\text{g}}{\text{mL}}} = 580 \text{ mL}$$

18.

$$m = 15.7 \text{ kg} \cdot \frac{1000 \text{ g}}{1 \text{ kg}} = 15,700 \text{ g}$$

$$\rho = 5.32 \; \frac{\text{g}}{\text{cm}^3}$$

$$\rho = \frac{m}{V} \rightarrow V = \frac{m}{\rho} = \frac{15,700 \text{ g}}{5.32 \; \frac{\text{g}}{\text{cm}^3}} = 2950 \text{ cm}^3$$

$$2950 \text{ cm}^3 \cdot \frac{1 \text{ m}}{100 \text{ cm}} \cdot \frac{1 \text{ m}}{100 \text{ cm}} \cdot \frac{1 \text{ m}}{100 \text{ cm}} = 0.00295 \text{ m}^3$$

19.

$m = 32.1$ g

$V_1 = 23.35$ mL

$V_2 = 27.79$ mL

$V = V_2 - V_1 = 27.79 \text{ mL} - 23.35 \text{ mL} = 4.44 \text{ mL}$

$\rho = \dfrac{m}{V} = \dfrac{32.1 \text{ g}}{4.44 \text{ mL}} = 7.23 \dfrac{\text{g}}{\text{mL}} \cdot \dfrac{1 \text{ mL}}{1 \text{ cm}^3} = 7.23 \dfrac{\text{g}}{\text{cm}^3}$

20.

$h = 34.5 \text{ in} \cdot \dfrac{2.54 \text{ cm}}{1 \text{ in}} \cdot \dfrac{1 \text{ m}}{100 \text{ cm}} = 0.876 \text{ m}$

$d = 24 \text{ in} \cdot \dfrac{2.54 \text{ cm}}{1 \text{ in}} \cdot \dfrac{1 \text{ m}}{100 \text{ cm}} = 0.6096 \text{ m}$

$r = \dfrac{d}{2} = \dfrac{0.6096 \text{ m}}{2} = 0.3048 \text{ m}$

$V = \pi r^2 h = 3.142 \cdot (0.3048 \text{ m})^2 \cdot 0.876 \text{ m} = 0.256 \text{ m}^3$

$\rho = 810 \dfrac{\text{kg}}{\text{m}^3}$

$\rho = \dfrac{m}{V} \rightarrow m = \rho V = 810 \dfrac{\text{kg}}{\text{m}^3} \cdot 0.256 \text{ m}^3 = 207 \text{ kg}$

Rounding this to 2 sig digs, we have 210 kg.

21.

$l = 2.1 \text{ cm} \cdot \dfrac{1 \text{ m}}{100 \text{ cm}} = 0.021 \text{ m}$

$w = 3.5 \text{ cm} \cdot \dfrac{1 \text{ m}}{100 \text{ cm}} = 0.035 \text{ m}$

$m = 94.5 \text{ g} \cdot \dfrac{1 \text{ kg}}{1000 \text{ g}} = 0.0945 \text{ kg}$

$V = lwh$

$\rho = 7830 \dfrac{\text{kg}}{\text{m}^3}$

$\rho = \dfrac{m}{V} = \dfrac{m}{lwh} \rightarrow h = \dfrac{m}{lw\rho} = \dfrac{0.0945 \text{ kg}}{0.021 \text{ m} \cdot 0.035 \text{ m} \cdot 7830 \dfrac{\text{kg}}{\text{m}^3}} = 0.016 \text{ m} \cdot \dfrac{100 \text{ cm}}{1 \text{ m}} = 1.6 \text{ cm}$

22.

$m = 306 \text{ g} \cdot \dfrac{1 \text{ kg}}{1000 \text{ g}} = 0.306 \text{ kg}$

$V = 22.5 \text{ mL} \cdot \dfrac{1 \text{ cm}^3}{1 \text{ mL}} \cdot \dfrac{1 \text{ m}}{100 \text{ cm}} \cdot \dfrac{1 \text{ m}}{100 \text{ cm}} \cdot \dfrac{1 \text{ m}}{100 \text{ cm}} = 0.0000225 \text{ m}^3$

$\rho = \dfrac{m}{V} = \dfrac{0.306 \text{ kg}}{0.0000225 \text{ m}^3} = 13,600 \dfrac{\text{kg}}{\text{m}^3}$

23.

$$V = 3.0 \times 10^6 \text{ gal} \cdot \frac{3.785 \text{ L}}{1 \text{ gal}} \cdot \frac{1 \text{ m}^3}{1000 \text{ L}} = 11,355 \text{ m}^3$$

$$\rho = 0.998 \frac{\text{g}}{\text{cm}^3} \cdot \frac{1 \text{ kg}}{1000 \text{ g}} \cdot \frac{100 \text{ cm}}{1 \text{ m}} \cdot \frac{100 \text{ cm}}{1 \text{ m}} \cdot \frac{100 \text{ cm}}{1 \text{ m}} = 998 \frac{\text{kg}}{\text{m}^3}$$

$$\rho = \frac{m}{V} \rightarrow m = \rho V = 998 \frac{\text{kg}}{\text{m}^3} \cdot 11,355 \text{ m}^3 = 11,000,000 \text{ kg}$$

24.

$$l = 45 \text{ ft} \cdot \frac{0.3048 \text{ m}}{1 \text{ ft}} = 13.7 \text{ m}$$

$$d = 2.0 \text{ ft}$$

$$r = \frac{d}{2} = \frac{2.0 \text{ ft}}{2} = 1.0 \text{ ft} \cdot \frac{0.3048 \text{ m}}{1 \text{ ft}} = 0.305 \text{ m}$$

$$\rho = 160 \frac{\text{kg}}{\text{m}^3}$$

$$V = \pi r^2 l = 3.14 \cdot (0.305 \text{ m})^2 \cdot 13.7 \text{ m} = 4.00 \text{ m}^3$$

$$\rho = \frac{m}{V} \rightarrow m = \rho V = 160 \frac{\text{kg}}{\text{m}^3} \cdot 4.00 \text{ m}^3 = 640 \text{ kg}$$

For nine logs:

$$640 \text{ kg} \cdot 9 = 5800 \text{ kg}$$

---

26. a.

$$73.2 \text{ g Cu} \cdot \frac{1 \text{ mol}}{63.55 \text{ g}} \cdot \frac{6.022 \times 10^{23} \text{ particles}}{\text{mol}} = 6.94 \times 10^{23} \text{ particles (atoms)}$$

---

26. b.

$$1.35 \text{ mol Na} \cdot \frac{6.022 \times 10^{23} \text{ particles}}{\text{mol}} = 8.13 \times 10^{23} \text{ particles (atoms)}$$

---

26. c.

$$1.5000 \text{ kg W} \cdot \frac{1000 \text{ g}}{1 \text{ kg}} \cdot \frac{\text{mol}}{183.85 \text{ g}} \cdot \frac{6.0221 \times 10^{23} \text{ particles}}{\text{mol}} = 4.9133 \times 10^{24} \text{ particles (atoms)}$$

---

27. a.

$$6.022 \times 10^{23} \text{ atoms K} \cdot \frac{\text{mol}}{6.022 \times 10^{23} \text{ atoms}} \cdot \frac{39.098 \text{ g}}{\text{mol}} = 39.10 \text{ g}$$

(4 sig digs because the given value has 4)

---

27. b.

$$100 \text{ atoms Au} \cdot \frac{\text{mol}}{6.022 \times 10^{23} \text{ atoms}} \cdot \frac{196.9665 \text{ g}}{\text{mol}} = 3 \times 10^{-20} \text{ g}$$

(1 sig dig because the given value has 1)

---

27. c.

$$0.00100 \text{ mol Xe} \cdot \frac{131.29 \text{ g}}{\text{mol}} = 0.131 \text{ g}$$

(3 sig digs because the given value has 3)

---

27. d.

$$2.0 \text{ mol Li} \cdot \frac{6.941 \text{ g}}{\text{mol}} = 14 \text{ g}$$

(2 sig digs because the given value has 2)

---

27. e.

$$4.2120 \text{ mol Br} \cdot \frac{79.904 \text{ g}}{\text{mol}} = 336.56 \text{ g}$$

(5 sig digs because the given value has 5)

---

27. f.

$$7.422 \times 10^{22} \text{ atoms Pt} \cdot \frac{\text{mol}}{6.022 \times 10^{23} \text{ atoms}} \cdot \frac{195.08 \text{ g}}{\text{mol}} = 24.04 \text{ g}$$

(4 sig digs because the given value has 4)

---

28. a.

$$25 \text{ g Ca(OH)}_2 \cdot \frac{\text{mol}}{74.09 \text{ g}} = 0.34 \text{ mol}$$

---

28. b.

$$286.25 \text{ g Al}_2(\text{CrO}_4)_3 \cdot \frac{\text{mol}}{401.944 \text{ g}} = 0.71216 \text{ mol}$$

---

28. c

$$2.111 \text{ kg KCl} \cdot \frac{1000 \text{ g}}{1 \text{ kg}} \cdot \frac{\text{mol}}{74.551 \text{ g}} = 28.32 \text{ mol}$$

---

28. d.

$$47.50 \text{ g LiClO}_3 \cdot \frac{\text{mol}}{90.39 \text{ g}} = 0.5255 \text{ mol}$$

---

28. e.

$$10.0 \text{ g O}_2 \cdot \frac{\text{mol}}{32.0 \text{ g}} = 0.313 \text{ mol}$$

---

28. f.

$$1.00 \text{ mg C}_{14}\text{H}_{18}\text{N}_2\text{O}_5 \cdot \frac{1 \text{ g}}{1000 \text{ mg}} \cdot \frac{\text{mol}}{294.307 \text{ g}} = 3.40 \times 10^{-6} \text{ mol}$$

---

29. a. For all items in problem 29, sig digs are determined as in problem 12.

$$1 \cdot 14.0067 \frac{\text{g}}{\text{mol}} = 14.0067 \frac{\text{g}}{\text{mol}}$$

$$3 \cdot 1.0079 \frac{\text{g}}{\text{mol}} = 3.0237 \frac{\text{g}}{\text{mol}}$$

14.0067 g/mol
+ 3.0237 g/mol

17.0304 g/mol

---

29. b.

$$1 \cdot 12.011 \frac{\text{g}}{\text{mol}} = 12.011 \frac{\text{g}}{\text{mol}}$$

$$2 \cdot 15.9994 \frac{\text{g}}{\text{mol}} = 31.9988 \frac{\text{g}}{\text{mol}}$$

12.011 g/mol
+ 31.9988 g/mol

44.010 g/mol

---

29. c.

$$2 \cdot 35.4527 \frac{\text{g}}{\text{mol}} = 70.9054 \frac{\text{g}}{\text{mol}}$$

---

29. d.

$$1 \cdot 63.546 \frac{\text{g}}{\text{mol}} = 63.546 \frac{\text{g}}{\text{mol}}$$

$$1 \cdot 32.066 \frac{\text{g}}{\text{mol}} = 32.066 \frac{\text{g}}{\text{mol}}$$

$$4 \cdot 15.9994 \frac{\text{g}}{\text{mol}} = 63.9976 \frac{\text{g}}{\text{mol}}$$

63.546 g/mol
32.066 g/mol
+ 63.9976 g/mol

159.610 g/mol

---

29. e.

$$1 \cdot 40.078 \frac{\text{g}}{\text{mol}} = 40.078 \frac{\text{g}}{\text{mol}}$$

$$2 \cdot 14.0067 \frac{\text{g}}{\text{mol}} = 28.0134 \frac{\text{g}}{\text{mol}}$$

$$4 \cdot 15.9994 \frac{\text{g}}{\text{mol}} = 63.9976 \frac{\text{g}}{\text{mol}}$$

40.078 g/mol
28.0134 g/mol
+ 63.9976 g/mol

132.089 g/mol

---

29. f.

$$12 \cdot 12.011 \ \frac{g}{mol} = 144.132 \ \frac{g}{mol}$$

$$22 \cdot 1.0079 \ \frac{g}{mol} = 22.174 \ \frac{g}{mol}$$

$$11 \cdot 15.9994 \ \frac{g}{mol} = 175.993 \ \frac{g}{mol}$$

144.132 g/mol
22.174 g/mol
+ 175.993 g/mol
342.30 g/mol

29. g.

$$2 \cdot 12.011 \ \frac{g}{mol} = 24.022 \ \frac{g}{mol}$$

$$6 \cdot 1.0079 \ \frac{g}{mol} = 6.0474 \ \frac{g}{mol}$$

$$1 \cdot 15.9994 \ \frac{g}{mol} = 15.9994 \ \frac{g}{mol}$$

24.022 g/mol
6.0474 g/mol
+ 15.9994 g/mol
46.069 g/mol

29. h.

$$3 \cdot 12.011 \ \frac{g}{mol} = 36.033 \ \frac{g}{mol}$$

$$8 \cdot 1.0079 \ \frac{g}{mol} = 8.0632 \ \frac{g}{mol}$$

36.033 g/mol
+ 8.0632 g/mol
44.096 g/mol

29. i.

$$1 \cdot 28.0855 \ \frac{g}{mol} = 28.0855 \ \frac{g}{mol}$$

$$2 \cdot 15.9994 \ \frac{g}{mol} = 31.9988 \ \frac{g}{mol}$$

28.0855 g/mol
+ 31.9988 g/mol
60.0843 g/mol

30. a. For all items in problem 30, sig digs are determined as in problem 12.

$1 \cdot 24.3050 \ u = 24.3050 \ u$

$2 \cdot 35.4527 \ u = 70.9054 \ u$

24.3050 u
+ 70.9054 u
95.2104 u

30. b.

$1 \cdot 40.078 \ u = 40.078 \ u$

$2 \cdot 14.0067 \ u = 28.0134 \ u$

$6 \cdot 15.9994 \ u = 95.9964 \ u$

40.078 u
28.0134 u
+ 95.9964 u
164.088 u

30. c.

$1 \cdot 32.066 \ u = 32.066 \ u$

$4 \cdot 15.9994 \ u = 63.9976 \ u$

32.066 u
+ 63.9976 u
96.064 u

30. d.

$$63.546 \text{ u}$$

$1 \cdot 63.546 \text{ u} = 63.546 \text{ u}$ 32.066 u

$1 \cdot 32.066 \text{ u} = 32.066 \text{ u}$ $+\ \ 63.9976 \text{ u}$

$4 \cdot 15.9994 \text{ u} = 63.9976 \text{ u}$ 159.610 u

---

30. e.

$$10.811 \text{ u}$$

$1 \cdot 10.811 \text{ u} = 10.811 \text{ u}$ $+\ \ 56.9952 \text{ u}$

$3 \cdot 18.9984 \text{ u} = 56.9952 \text{ u}$ 67.806 u

---

30. f.

$$12.011 \text{ u}$$

$1 \cdot 12.011 \text{ u} = 12.011 \text{ u}$ $+\ \ 141.811 \text{ u}$

$4 \cdot 35.4527 \text{ u} = 141.811 \text{ u}$ 153.822 u

31.

$$2.25 \text{ mol AgNO}_3 \cdot \frac{169.8731 \text{ g}}{\text{mol}} = 382 \text{ g}$$

(3 sig digs because the given value has 3)

---

32. a.

$$2.25 \text{ kg CCl}_4 \cdot \frac{1000 \text{ g}}{\text{kg}} \cdot \frac{\text{mol}}{153.822 \text{ g}} = 14.6 \text{ mol}$$

(3 sig digs because the given value has 3)

---

32. b.

$$14.63 \text{ mol} \cdot \frac{6.022 \times 10^{23} \text{ particles}}{\text{mol}} = 8.81 \times 10^{24} \text{ particles}$$

For CCl_4, particles = molecules, and each molecule contains 1 carbon atom, giving 8.81×10^{24} carbon atoms. (Computing moles in step 1 was an intermediate calculation for this one, thus an extra digit was retained to get 14.63.)

---

32. c.

From Table 2.6, 1.078% of carbon atoms are carbon-13. This gives $8.81 \times 10^{24} \cdot 0.01078 = 9.50 \times 10^{22}$ atoms of carbon-13.

---

33. a.

$$1.00 \text{ gal} \cdot \frac{3.785 \text{ L}}{1 \text{ gal}} \cdot \frac{1000 \text{ mL}}{1 \text{ L}} = 3785 \text{ mL}$$

$$\rho = \frac{m}{V} \rightarrow m = \rho V = 1.000 \ \frac{g}{mL} \cdot 3785 \text{ mL} = 3785 \text{ g}$$

$$3785 \text{ g} \cdot \frac{\text{mol}}{18.02 \text{ g}} = 2.10 \times 10^2 \text{ mol}$$

(Result written in scientific notation in order to show 3 sig digs.)

33. b.

$$210.0 \text{ mol} \cdot \frac{6.022 \times 10^{23} \text{ particles}}{\text{mol}} = 1.264 \times 10^{26} \text{ particles}$$

Particles = molecules, and there are two H atoms in each molecule. Doubling and rounding to 3 sig digs gives 2.53×10^{26} H atoms.

33. c.

From Table 2.6, H-2 is 0.0115% of all hydrogen. $2.53 \times 10^{26} \cdot 0.000115 = 2.91 \times 10^{22}$ atoms of H-2.

Chapter 3

4.

$$\lambda = 543 \text{ nm} \cdot \frac{1 \text{ m}}{1 \times 10^9 \text{ nm}} = 5.43 \times 10^{-7} \text{ m}$$

$$E = \frac{hv}{\lambda} = \frac{6.626 \times 10^{-34} \text{ J} \cdot \text{s} \cdot 2.9979 \times 10^8 \frac{\text{m}}{\text{s}}}{5.43 \times 10^{-7} \text{ m}} = 3.66 \times 10^{-19} \text{ J}$$

5.

With 5 sig digs in the given value, we need to use constants from Appendix A with at least 5 sig digs.

$$E = 2.2718 \times 10^{-19} \text{ J}$$

$$E = \frac{hv}{\lambda} \rightarrow \lambda = \frac{hv}{E} = \frac{6.62607 \times 10^{-34} \text{ J} \cdot \text{s} \cdot 2.9979 \times 10^8 \frac{\text{m}}{\text{s}}}{2.2718 \times 10^{-19} \text{ J}} = 8.7439 \times 10^{-7} \text{ m}$$

$$8.7439 \times 10^{-7} \text{ m} \cdot \frac{1 \times 10^9 \text{ nm}}{1 \text{ m}} = 874.39 \text{ nm}$$

The result has 5 sig digs because the given value has 5.

6.

From Figure 3.6, $E = 2.18 \times 10^{-18}$ J

$$E = \frac{hv}{\lambda} \rightarrow \lambda = \frac{hv}{E} = \frac{6.626 \times 10^{-34} \text{ J} \cdot \text{s} \cdot 2.9979 \times 10^8 \frac{\text{m}}{\text{s}}}{2.18 \times 10^{-18} \text{ J}} = 9.11 \times 10^{-8} \text{ m} \cdot \frac{10^9 \text{ nm}}{1 \text{ m}} = 91.1 \text{ nm}$$

The result has 3 sig digs because the given value has 3.

7.

The visible spectrum is the Balmer series. From Figure 3.8,

$$\lambda_1 = 410 \text{ nm} \cdot \frac{1 \text{ m}}{1 \times 10^9 \text{ nm}} = 4.1 \times 10^{-7} \text{ m}$$

$$\lambda_2 = 434 \text{ nm} \cdot \frac{1 \text{ m}}{1 \times 10^9 \text{ nm}} = 4.34 \times 10^{-7} \text{ m}$$

$$\lambda_3 = 486 \text{ nm} \cdot \frac{1 \text{ m}}{1 \times 10^9 \text{ nm}} = 4.86 \times 10^{-7} \text{ m}$$

$$\lambda_4 = 656 \text{ nm} \cdot \frac{1 \text{ m}}{1 \times 10^9 \text{ nm}} = 6.56 \times 10^{-7} \text{ m}$$

$$E_1 = \frac{h\nu}{\lambda} = \frac{6.626 \times 10^{-34} \text{ J} \cdot \text{s} \cdot 2.9979 \times 10^8 \frac{\text{m}}{\text{s}}}{4.1 \times 10^{-7} \text{ m}} = 4.8 \times 10^{-19} \text{ J} \left(2 \text{ sig digs because 410 has 2}\right)$$

$$E_2 = \frac{h\nu}{\lambda} = \frac{6.626 \times 10^{-34} \text{ J} \cdot \text{s} \cdot 2.9979 \times 10^8 \frac{\text{m}}{\text{s}}}{4.34 \times 10^{-7} \text{ m}} = 4.58 \times 10^{-19} \text{ J} \left(3 \text{ sig digs because 434 has 3}\right)$$

$$E_3 = \frac{h\nu}{\lambda} = \frac{6.626 \times 10^{-34} \text{ J} \cdot \text{s} \cdot 2.9979 \times 10^8 \frac{\text{m}}{\text{s}}}{4.86 \times 10^{-7} \text{ m}} = 4.09 \times 10^{-19} \text{ J} \left(3 \text{ sig digs because 486 has 3}\right)$$

$$E_4 = \frac{h\nu}{\lambda} = \frac{6.626 \times 10^{-34} \text{ J} \cdot \text{s} \cdot 2.9979 \times 10^8 \frac{\text{m}}{\text{s}}}{6.56 \times 10^{-7} \text{ m}} = 3.03 \times 10^{-19} \text{ J} \left(3 \text{ sig digs because 656 has 3}\right)$$

24.

Assuming a 100-gram sample:

$$38.7 \text{ g C} \cdot \frac{\text{mol}}{12.011 \text{ g}} = 3.222 \text{ mol C} \quad (3.222/3.222 = 1)$$

$$9.7 \text{ g H} \cdot \frac{\text{mol}}{1.0079 \text{ g}} = 9.62 \text{ mol H} \quad (9.62/3.222 = 2.99 \approx 3)$$

$$51.6 \text{ g O} \cdot \frac{\text{mol}}{15.9994 \text{ g}} = 3.225 \text{ mol O} \quad (3.225/3.222 \approx 1)$$

These ratios give an empirical formula of CH_3O.

$$1 \cdot 12.011 \text{ u} + 3 \cdot 1.0079 \text{ u} + 1 \cdot 15.9994 \text{ u} = 31.0 \text{ u}$$

$$62.1 \text{ u} / 31.0 \text{ u} \approx 2$$

Applying the factor of 2 to the empirical formula gives a molecular formula of $C_2H_6O_2$.

---

25. a.

$$\text{C:} \quad \frac{43.910 \text{ g}}{47.593 \text{ g}} = 0.92261 = 92.261\% \text{ C}$$

$$\text{H:} \quad \frac{3.683 \text{ g}}{47.593 \text{ g}} = 0.07739 = 7.739\% \text{ H}$$

25. b.

Assuming a 100-gram sample:

$$92.261 \text{ g C} \cdot \frac{\text{mol}}{12.011 \text{ g}} = 7.68 \text{ mol} \quad (7.68/7.68 = 1)$$

$$7.739 \text{ g H} \cdot \frac{\text{mol}}{1.0079 \text{ g}} = 7.68 \text{ mol} \quad (7.68/7.68 = 1)$$

These ratios give an empirical formula of CH.

25. c.

$1 \cdot 12.011 \text{ u} + 1 \cdot 1.0079 \text{ u} = 13.019 \text{ u}$

$78.11 \text{ u}/13.019 \text{ u} = 6$

Applying the factor of 6 to the empirical formula gives a molecular formula of C_6H_6.

26. a.

$1 \cdot 22.9898 \text{ u} + 1 \cdot 1.0079 \text{ u} + 1 \cdot 12.011 \text{ u} + 3 \cdot 15.9994 \text{ u} = 84.007 \text{ u}$

$22.9898/84.007 = 0.27367 \rightarrow 27.367\% \text{ Na}$

$1.0079/84.007 = 0.011998 \rightarrow 1.1998\% \text{ H}$

$12.011/84.007 = 0.14298 \rightarrow 14.298\% \text{ C}$

$(3 \cdot 15.9994)/84.007 = 0.57136 \rightarrow 57.136\% \text{ O}$

The sig digs in the formula mass are limited by the decimals in the mass of carbon. When this value came out with 5 sig digs, all percentages after that had 5 sig digs.

26. b.

$2 \cdot 22.9898 \text{ u} + 1 \cdot 15.9994 \text{ u} = 61.9790 \text{ u}$

$(2 \cdot 22.9898)/61.9790 = 0.741858 \rightarrow 74.1858\% \text{ Na}$

$(1 \cdot 15.9994)/61.9790 = 0.258142 \rightarrow 25.8142\% \text{ O}$

26. c.

$2 \cdot 55.847 \text{ u} + 3 \cdot 15.9994 \text{ u} = 159.692 \text{ u}$

$(2 \cdot 55.847)/159.692 = 0.69943 \rightarrow 69.943\% \text{ Fe}$

$(3 \cdot 15.9994)/159.692 = 0.300567 \rightarrow 30.0567\% \text{ O}$

The sig digs in the formula mass are limited by the decimals in the mass of iron. The atomic mass of iron also limits the iron percentage to 5 sig digs.

26. d.

$1 \cdot 107.8682 \text{ u} + 1 \cdot 14.0067 \text{ u} + 3 \cdot 15.9994 \text{ u} = 169.8731 \text{ u}$

$(1 \cdot 107.8682)/169.8731 = 0.6349928 \rightarrow 63.49928\% \text{ Ag}$

$(1 \cdot 14.0067)/169.8731 = 0.0824539 \rightarrow 8.24539\% \text{ N}$

$(3 \cdot 15.9994)/169.8731 = 0.282553 \rightarrow 28.2553\% \text{ O}$

26. e.

$1 \cdot 40.078 \, u + 4 \cdot 12.011 \, u + 6 \cdot 1.0079 \, u + 4 \cdot 15.9994 \, u = 158.167 \, u$

$(1 \cdot 40.078)/158.167 = 0.25339 \rightarrow 25.339\% \; Ca$

$(4 \cdot 12.011)/158.167 = 0.30375 \rightarrow 30.375\% \; C$

$(6 \cdot 1.0079)/158.167 = 0.038234 \rightarrow 3.8234\% \; H$

$(4 \cdot 15.9994)/158.167 = 0.404620 \rightarrow 40.4620\% \; O$

The sig digs in the formula mass are limited to the third decimal by Ca and C. In the results, the sig digs for Ca, C, and H are limited by the atomic masses to 5. O has 6 sig digs in the mass, thus 6 in the result.

26. f.

$9 \cdot 12.011 \, u + 8 \cdot 1.0079 \, u + 4 \cdot 15.9994 \, u = 180.160 \, u$

$(9 \cdot 12.011)/180.160 = 0.60002 \rightarrow 60.002\% \; C$

$(8 \cdot 1.0079)/180.160 = 0.044756 \rightarrow 4.4756\% \; H$

$(4 \cdot 15.9994)/180.160 = 0.355226 \rightarrow 35.5226\% \; O$

27.

$1 \cdot 65.39 \, u + 1 \cdot 32.066 \, u + 4 \cdot 15.9994 \, u + 14 \cdot 1.0079 \, u + 7 \cdot 15.9994 \, u = 287.56 \, u$

$14 \cdot 1.0079 \, u + 7 \cdot 15.9994 \, u = 126.1064 \, u$

$126.1064 / 287.56 = 0.43854 \; (43.854\%)$

28.

Assuming a 100-gram sample:

$22.65 \, g \; S \cdot \dfrac{mol}{32.066 \, g} = 0.7064 \; mol \quad (0.7064/0.7064 = 1)$

$32.38 \, g \; Na \cdot \dfrac{mol}{22.9898 \, g} = 1.408 \; mol \quad (1.408/0.7064 \approx 2)$

$44.99 \, g \; O \cdot \dfrac{mol}{15.9994 \, g} = 2.812 \; mol \quad (2.812/0.7064 \approx 4)$

These ratios give an empirical formula of SNa_2O_4. As we will see later, this is actually Na_2SO_4.

29.

$1 \cdot 12.011 \, u + 2 \cdot 1.0079 \, u + 1 \cdot 15.9994 \, u = 30.026 \, u$

$120.12 / 30.026 \approx 4$

Applying the factor of 4 to the empirical formula gives a molecular formula of $C_4H_8O_4$.

30. a.

Assuming a 100-gram sample:

$$49.5 \text{ g C} \cdot \frac{\text{mol}}{12.011} = 4.12 \text{ mol} \quad (4.12/1.03 = 4)$$

$$5.15 \text{ g H} \cdot \frac{\text{mol}}{1.0079} = 5.11 \text{ mol} \quad (5.11/1.03 \approx 5)$$

$$28.9 \text{ g N} \cdot \frac{\text{mol}}{14.0067} = 2.06 \text{ mol} \quad (2.06/1.03 = 2)$$

$$16.5 \text{ g O} \cdot \frac{\text{mol}}{15.9994} = 1.03 \text{ mol} \quad (1.03/1.03 = 1)$$

These ratios give an empirical formula of $C_4H_5N_2O$. We get the multiplier factor from the empirical formula mass:

$$4 \cdot 12.011 \text{ u} + 5 \cdot 1.0079 \text{ u} + 2 \cdot 14.0067 \text{ u} + 1 \cdot 15.9994 \text{ u} = 97.096 \text{ u}$$

$$195/97.096 \approx 2$$

This factor gives a molecular formula of $C_8H_{10}N_4O_2$.

30. b.

Assuming a 100-gram sample:

$$75.69 \text{ g C} \cdot \frac{\text{mol}}{12.011} = 6.302 \text{ mol} \quad (6.302/0.9694 \approx 6.5) \to (6.5 \cdot 2 = 13)$$

$$8.80 \text{ g H} \cdot \frac{\text{mol}}{1.0079} = 8.73 \text{ mol} \quad (8.73/0.9694 \approx 9) \to (9 \cdot 2 = 18)$$

$$15.51 \text{ g O} \cdot \frac{\text{mol}}{15.9994} = 0.9694 \text{ mol} \quad (0.9694/0.9694 = 1) \to (1 \cdot 2 = 2)$$

The values are all doubled to obtain whole number ratios. These ratios give an empirical formula of $C_{13}H_{18}O_2$. We get the multiplier factor from the empirical formula mass:

$$13 \cdot 12.011 \text{ u} + 18 \cdot 1.0079 \text{ u} + 2 \cdot 15.9994 \text{ u} = 206.284 \text{ u}$$

$$206/206.284 \approx 1$$

The factor of 1 indicates that the empirical and molecular formulas are the same.

30. c.

Assuming a 100-gram sample:

$$81.71 \text{ g C} \cdot \frac{\text{mol}}{12.011} = 6.803 \text{ mol} \quad (6.803/6.803 = 1) \to (1 \cdot 3 = 3)$$

$$18.29 \text{ g H} \cdot \frac{\text{mol}}{1.0079} = 18.15 \text{ mol} \quad (18.15/6.803 \approx 2.668) \to (2.668 \cdot 3 \approx 8)$$

The values are all tripled to obtain whole number ratios. These ratios give an empirical formula of C_3H_8. We get the multiplier factor from the empirical formula mass:

$3 \cdot 12.011 \, u + 8 \cdot 1.0079 \, u = 44.096 \, u$

$44.096 / 44.096 = 1$

The factor of 1 indicates that the empirical and molecular formulas are the same.

30. d.

Assuming a 100-gram sample:

$$57.14 \, g \, C \cdot \frac{mol}{12.011} = 4.757 \, mol \quad (4.757/0.680 \approx 7) \rightarrow (7 \cdot 2 = 14)$$

$$6.16 \, g \, H \cdot \frac{mol}{1.0079} = 6.11 \, mol \quad (6.11/0.680 \approx 9) \rightarrow (9 \cdot 2 = 18)$$

$$9.52 \, g \, N \cdot \frac{mol}{14.0067} = 0.680 \, mol \quad (0.680/0.680 = 1) \rightarrow (1 \cdot 2 = 2)$$

$$27.18 \, g \, O \cdot \frac{mol}{15.9994} = 1.699 \, mol \quad (1.699/0.680 = 2.5) \rightarrow (2.5 \cdot 2 = 5)$$

The values are all doubled to obtain whole number ratios. These ratios give an empirical formula of $C_{14}H_{18}N_2O_5$. We get the multiplier factor from the empirical formula mass:

$14 \cdot 12.011 \, u + 18 \cdot 1.0079 \, u + 2 \cdot 14.0067 \, u + 5 \cdot 15.9994 \, u = 294.31 \, u$

$294.302 / 294.31 \approx 1$

The factor of 1 indicates that the empirical and molecular formulas are the same.

30. e.

Assuming a 100-gram sample:

$$92.26 \, g \, C \cdot \frac{mol}{12.011} = 7.681 \, mol \quad (7.681/7.68 \approx 1)$$

$$7.74 \, g \, H \cdot \frac{mol}{1.0079} = 7.68 \, mol \quad (7.68/7.68 = 1)$$

These ratios give an empirical formula of CH. We get the multiplier factor from the empirical formula mass:

$1 \cdot 12.011 \, u + 1 \cdot 1.0079 \, u = 13.019 \, u$

$26.038 / 13.019 = 2$

This factor gives a molecular formula of C_2H_2.

31.

$9.581/10.5 = 0.912 \rightarrow 91.2\%$ C

$0.919/10.5 = 0.0875 \rightarrow 8.75\%$ H

Assuming a 100-gram sample:

$$91.2 \text{ g C} \cdot \frac{\text{mol}}{12.011 \text{ g}} = 7.59 \text{ mol} \quad (7.59/7.59 = 1) \rightarrow (1 \cdot 7 = 7)$$

$$8.75 \text{ g H} \cdot \frac{\text{mol}}{1.0079 \text{ g}} = 8.68 \text{ mol} \quad (8.68/7.59 \approx 1.14) \rightarrow (1.14 \cdot 7 = 8)$$

The values are multiplied by 7 to obtain whole number ratios. These ratios give an empirical formula of C_7H_8. We get the multiplier factor from the empirical formula mass:

$7 \cdot 12.011 \text{ u} + 8 \cdot 1.0079 \text{ u} = 92.140 \text{ u}$

$92.140/92.140 = 1$

The factor of 1 indicates that the empirical and molecular formulas are the same.

32.

$1.0079 \text{ u} \cdot 2 + 32.066 \text{ u} \cdot 1 + 15.9994 \text{ u} \cdot 4 = 98.079 \text{ u}$ (formula mass)

$98.079 \dfrac{\text{g}}{\text{mol}}$ (molar mass)

33.

$$125.0 \text{ g HClO}_3 \cdot \frac{\text{mol}}{84.459 \text{ g}} \cdot \frac{6.022 \times 10^{23} \text{ particles}}{\text{mol}} = 8.913 \times 10^{23} \text{ particles}$$
$$\text{(molecules)}$$

34.

$l = 50.0$ m

$w = 25.0$ m

$h = 2.00$ m

$$\rho = 0.9970 \frac{\text{g}}{\text{cm}^3} \cdot \frac{1 \text{ kg}}{1000 \text{ g}} \cdot \frac{100 \text{ cm}}{1 \text{ m}} \cdot \frac{100 \text{ cm}}{1 \text{ m}} \cdot \frac{100 \text{ cm}}{1 \text{ m}} = 997.0 \frac{\text{kg}}{\text{m}^3}$$

$V = lwh = 50.0 \text{ m} \cdot 25.0 \text{ m} \cdot 2.00 \text{ m} = 2500 \text{ m}^3$

$$\rho = \frac{m}{V} \rightarrow m = \rho V = 997.0 \frac{\text{kg}}{\text{m}^3} \cdot 2500 \text{ m}^3 = 2.49 \times 10^6 \text{ kg}$$

35.

$$1{,}000{,}000 \text{ particles} \cdot \frac{\text{mol}}{6.022 \times 10^{23} \text{ particles}} \cdot \frac{207.2 \text{ g}}{\text{mol}} = 3.441 \times 10^{-16} \text{ g}$$

$$3.441 \times 10^{-16} \text{ g} \cdot \frac{10^6 \text{ µg}}{1 \text{ g}} = 3.441 \times 10^{-10} \text{ µg}$$

36.

$$V = 1.000 \text{ pt} \cdot \frac{473.176 \text{ mL}}{1 \text{ pt}} \cdot \frac{1 \text{ cm}^3}{1 \text{ mL}} = 473.176 \text{ cm}^3$$

$$\rho = 1.049 \ \frac{\text{g}}{\text{cm}^3}$$

$$\rho = \frac{m}{V} \rightarrow m = \rho V = 1.049 \ \frac{\text{g}}{\text{cm}^3} \cdot 473.176 \text{ cm}^3 = 496.36 \text{ g}$$

$$496.36 \text{ g} \cdot \frac{\text{mol}}{60.052 \text{ g}} \cdot \frac{6.022 \times 10^{23} \text{ particles}}{\text{mol}} = 4.977 \times 10^{24} \text{ particles (molecules)}$$

37.

$$F_w = 20.0 \text{ lb} \cdot \frac{4.448 \text{ N}}{1 \text{ lb}} = 88.96 \text{ N}$$

$$m = \frac{F_w}{g} = \frac{88.96 \text{ N}}{9.80 \ \frac{\text{m}}{\text{s}^2}} = 9.078 \text{ kg} = 9078 \text{ g}$$

$$9078 \text{ g} \cdot \frac{1 \text{ mol}}{44.096 \text{ g}} = 205.9 \text{ mol}$$

$$205.9 \text{ mol} \cdot \frac{6.022 \times 10^{23} \text{ molecules}}{\text{mol}} = 1.24 \times 10^{26} \text{ molecules}$$

38.

$$\rho = 19.30 \ \frac{\text{g}}{\text{cm}^3}$$

$$3.000 \text{ mol} \cdot \frac{196.97 \text{ g}}{\text{mol}} = 590.91 \text{ g} = m$$

$$\rho = \frac{m}{V} \rightarrow V = \frac{m}{\rho} = \frac{590.91 \text{ g}}{19.30 \ \frac{\text{g}}{\text{cm}^3}} = 30.62 \text{ cm}^3$$

---

46. a.

$$13.96 \text{ mL} \cdot \frac{1 \text{ L}}{1000 \text{ mL}} \cdot \frac{1 \text{ gal}}{3.7854 \text{ L}} = 0.003688 \text{ gal}$$

---

46. b.

$$T_C = \frac{5}{9}(T_F - 32°) = \frac{5}{9}(-62.7°F - 32°) = -52.6°$$

$$T_K = T_C + 273.15 = -52.6 + 273.15 = 220.6 \text{ K}$$

---

46. c.

$$2.2901 \times 10^{-3} \text{ m}^3 \cdot \frac{100 \text{ cm}}{1 \text{ m}} \cdot \frac{100 \text{ cm}}{1 \text{ m}} \cdot \frac{100 \text{ cm}}{1 \text{ m}} \cdot \frac{1 \text{ in}}{2.54 \text{ cm}} \cdot \frac{1 \text{ in}}{2.54 \text{ cm}} \cdot \frac{1 \text{ in}}{2.54 \text{ cm}} = 139.75 \text{ in}^3$$

---

46. d.

$$1.2509\times10^4 \text{ nm}\cdot\frac{1\text{ m}}{10^9\text{ nm}}\cdot\frac{10^6\text{ }\mu\text{m}}{1\text{ m}}=12.509\text{ }\mu\text{m}$$

46. e.

$$130,005\text{ kPa}\cdot\frac{1000\text{ Pa}}{1\text{ kPa}}\cdot\frac{1\text{ bar}}{100,000\text{ Pa}}=1300.05\text{ bar}$$

46. f.

$$60,000\text{ mi}\cdot\frac{5280\text{ ft}}{1\text{ mi}}\cdot\frac{12\text{ in}}{1\text{ ft}}\cdot\frac{2.54\text{ cm}}{1\text{ in}}=10,000,000,000\text{ cm}$$

46. g.

$$T_F=\frac{9}{5}T_C+32°=\frac{9}{5}\cdot65.7°\text{C}+32°=150°\text{F}$$

To write this with 3 sig digs, we must write 1.50×10^2 °F.

46. h.

$$10,600\text{ mL}\cdot\frac{1\text{ L}}{1000\text{ mL}}=10.6\text{ L}$$

46. i.

$$23.17\text{ }\mu\text{s}\cdot\frac{1\text{ s}}{10^6\text{ }\mu\text{s}}=2.317\times10^{-5}\text{ s}$$

46. j.

$$T_C=\frac{5}{9}\left(T_F-32°\right)=\frac{5}{9}\left(98.6°\text{F}-32°\right)=37.0°$$

46. k.

$$1.600076\times10^{-9}\text{ km}\cdot\frac{1000\text{ m}}{1\text{ km}}\cdot\frac{100\text{ cm}}{1\text{ m}}=1.600076\times10^{-4}\text{ cm}$$

46. l.

$$1.002\times10^{-14}\text{ kg}\cdot\frac{1000\text{ g}}{1\text{ kg}}\cdot\frac{10^9\text{ ng}}{1\text{ g}}=0.01002\text{ ng}$$

Chapter 4

30.

$$\lambda = 94 \text{ nm} \cdot \frac{1 \text{ m}}{1 \times 10^9 \text{ nm}} = 9.4 \times 10^{-8} \text{ m}$$

$$E = \frac{hv}{\lambda} = \frac{6.626 \times 10^{-34} \text{ J} \cdot \text{s} \cdot 2.9979 \times 10^8 \frac{\text{m}}{\text{s}}}{9.4 \times 10^{-8} \text{ m}} = 2.11 \times 10^{-18} \text{ J} \cdot \frac{1 \text{ eV}}{1.60 \times 10^{-19} \text{ J}} = 13 \text{ eV}$$

31.

$$E = 5.09 \times 10^{-19} \text{ J}$$

$$E = \frac{hv}{\lambda} \rightarrow \lambda = \frac{hv}{E} = \frac{6.626 \times 10^{-34} \text{ J} \cdot \text{s} \cdot 2.9979 \times 10^8 \frac{\text{m}}{\text{s}}}{5.09 \times 10^{-19} \text{ J}} = 3.903 \times 10^{-7} \text{ m} \cdot \frac{1 \times 10^9 \text{ nm}}{1 \text{ m}} = 390 \text{ nm}$$

The result is stated with only 2 sig digs instead of three so we can see it in nanometers and compare it to the visible band of 700–400 nm. Doing so, the light is in the ultraviolet band and would thus not be visible.

34.

$$112 \text{ g CO}_2 \cdot \frac{\text{mol}}{44.01 \text{ g}} \cdot \frac{6.022 \times 10^{23} \text{ particles}}{\text{mol}} = 1.53 \times 10^{24} \text{ particles}$$

For CO_2, the particles are molecules and there is one atom of C in each molecule. Thus, this value is the number of carbon atoms.

35.

$$2 \cdot 1.0079 \text{ u} + 1 \cdot 32.066 \text{ u} + 4 \cdot 15.9994 \text{ u} = 98.079 \text{ u}$$

$$2 \cdot 1.0079 \text{ u} / 98.079 \text{ u} = 0.020553 \rightarrow 2.0553\% \text{ H}$$

$$1 \cdot 32.066 \text{ u} / 98.079 \text{ u} = 0.32694 \rightarrow 32.694\% \text{ S}$$

$$4 \cdot 15.9994 \text{ u} / 98.079 \text{ u} = 0.65251 \rightarrow 65.251\% \text{ O}$$

$$35.0 \text{ g H}_2\text{SO}_4 \cdot \frac{\text{mol}}{98.079 \text{ g}} \cdot \frac{6.022 \times 10^{23} \text{ particles}}{\text{mol}} = 2.15 \times 10^{23} \text{ particles}$$

Each particle is a molecule, and each molecule contains 2 H atoms, 1 S atom, and 4 O atoms. Thus, the numbers of atoms are 4.30×10^{23} H atoms, 2.15×10^{23} S atoms, and 8.60×10^{23} O atoms.

37.

$$3.00 \text{ mol CaBr}_2 \cdot \frac{199.89 \text{ g}}{\text{mol}} = 599.7 \text{ g}$$

$$40.078 \text{ u} / 199.89 \text{ u} = 0.2005 \text{ (proportion of calcium)}$$

$$0.2005 \cdot 599.7 \text{ g} = 1.20 \times 10^2 \text{ g}$$

The result is stated in scientific notation in order to show 3 sig digs.

38.

Assuming a 100-gram sample:

$$53.64 \text{ g Cl} \cdot \frac{\text{mol}}{35.4527 \text{ g}} = 1.513 \text{ mol} \quad (1.513/0.2522 = 6)$$

$$46.36 \text{ g W} \cdot \frac{\text{mol}}{183.85 \text{ g}} = 0.2522 \text{ mol} \quad (0.2522/0.2522 = 1)$$

The 6:1 ratio gives an empirical formula of WCl_6.

Chapter 5

20. a.

Be—F: 2.41, O—F: 0.54, C—F: 1.43

20. b.

F—F: 0, B—F: 1.94, S—O: 0.86

20. c.

O—Cl: 0.28, S—Cl: 0.58, C—P: 0.36

22.

$$100.00 \text{ g CaCO}_3 \cdot \frac{\text{mol}}{100.087 \text{ g}} \cdot \frac{6.02214 \times 10^{23} \text{ particles}}{\text{mol}} = 6.0169 \times 10^{23} \text{ particles}$$

In this ionic compound, the particles are formula units. There are 3 O atoms in each formula unit, so the number of O atoms is 1.8051×10^{24}.

23.

$1 \cdot 32.066 \text{ u} + 6 \cdot 18.9984 \text{ u} = 146.056 \text{ u}$

$1 \cdot 32.066 \text{ u} / 146.056 \text{ u} = 0.21955 \rightarrow 21.955\% \text{ S}$

$6 \cdot 18.9984 \text{ u} / 146.056 \text{ u} = 0.780457 \rightarrow 78.0457\% \text{ F}$

25.

$$\rho = \frac{m}{V} \rightarrow m = \rho V = 2.33 \frac{\text{g}}{\text{cm}^3} \cdot 1.000 \text{ cm}^3 = 2.33 \text{ g}$$

$$2.33 \text{ g MgCl}_2 \cdot \frac{\text{mol}}{95.2104 \text{ g}} \cdot \frac{6.022 \times 10^{23} \text{ particles}}{\text{mol}} = 1.47 \times 10^{22} \text{ particles}$$

In this ionic compound, the particles are formula units. There are 2 Cl atoms in each formula unit, so the number of Cl atoms is 2.94×10^{22}.

28.

Assuming a 100-gram sample:

$$12.84 \text{ g S} \cdot \frac{\text{mol}}{32.066 \text{ g}} = 0.4044 \text{ mol} \quad (0.4044 / 0.4011 \approx 1)$$

$$25.45 \text{ g Cu} \cdot \frac{\text{mol}}{63.546 \text{ g}} = 0.4011 \text{ mol} \quad (0.4011 / 0.4011 = 1)$$

$$25.63 \text{ g O} \cdot \frac{\text{mol}}{15.9994 \text{ g}} = 1.602 \text{ mol} \quad (1.602 / 0.4011 \approx 4)$$

$$36.07 \text{ g H}_2\text{O} \cdot \frac{\text{mol}}{18.0152 \text{ g}} = 2.002 \text{ mol} \quad (2.002 / 0.4011 \approx 5)$$

These ratios give an empirical formula of $CuSO_4 \cdot 5H_2O$. This is copper sulfate pentahydrate.

29.

$E = 3.73 \times 10^{-19}$ J

$$E = \frac{hv}{\lambda} \rightarrow \lambda = \frac{hv}{E} = \frac{6.626 \times 10^{-34} \text{ J} \cdot \text{s} \cdot 2.9979 \times 10^8 \ \frac{m}{s}}{3.73 \times 10^{-19} \text{ J}} = 5.33 \times 10^{-7} \text{ m}$$

$$5.33 \times 10^{-7} \text{ m} \cdot \frac{10^9 \text{ nm}}{1 \text{ m}} = 533 \text{ nm}$$

Chapter 6

27.

Formula: $C_3NO_2H_7$

$3 \cdot 12.011$ u $/ 89.094$ u $= 0.40444 \rightarrow 40.444\%$ C

$1 \cdot 14.0067$ u $/ 89.094$ u $= 0.15721 \rightarrow 15.721\%$ N

$2 \cdot 15.9994$ u $/ 89.094$ u $= 0.35916 \rightarrow 35.916\%$ O

$7 \cdot 1.0079$ u $/ 89.094$ u $= 0.079189 \rightarrow 7.9189\%$ H

28.

$$\lambda = 601 \text{ nm} \cdot \frac{1 \text{ m}}{1 \times 10^9 \text{ nm}} = 6.01 \times 10^{-7} \text{ m}$$

$$E = \frac{h\nu}{\lambda} = \frac{6.626 \times 10^{-34} \text{ J} \cdot \text{s} \cdot 2.9979 \times 10^8 \frac{\text{m}}{\text{s}}}{6.01 \times 10^{-7} \text{ m}} = 3.305 \times 10^{-19} \text{ J} \cdot \frac{1 \text{ eV}}{1.602 \times 10^{-19} \text{ J}} = 2.06 \text{ eV}$$

37. a.

$$65.6 \text{ g H}_2\text{O} \cdot \frac{\text{mol}}{18.02 \text{ g}} = 3.64 \text{ mol}$$

37. b.

$$1250 \text{ mg C}_6\text{H}_6\text{O}_6 \cdot \frac{\text{g}}{1000 \text{ mg}} \cdot \frac{\text{mol}}{174.110 \text{ g}} = 0.00718 \text{ mol}$$

37. c.

$$400 \text{ mg C}_9\text{H}_8\text{O}_4 \cdot \frac{\text{g}}{1000 \text{ mg}} \cdot \frac{\text{mol}}{180.160 \text{ g}} = 0.002 \text{ mol}$$

This result has been rounded to 1 sig dig since the given quantity (400 mg) has only 1 sig dig.

37. d.

$$14.0 \text{ kg KNO}_3 \cdot \frac{1000 \text{ g}}{1 \text{ kg}} \cdot \frac{\text{mol}}{101.103 \text{ g}} = 138 \text{ mol}$$

This result has been rounded to 3 sig digs since the given quantity (14.0 kg) has 3 sig digs.

37. e.

$$1050 \text{ g HClO}_4 \cdot \frac{\text{mol}}{100.4582 \text{ g}} = 10.5 \text{ mol}$$

This result has been rounded to 3 sig digs since the given quantity (1,050 g) has 3 sig digs.

37. f.

$$953.00 \text{ g HF} \cdot \frac{\text{mol}}{20.0063 \text{ g}} = 47.635 \text{ mol}$$

This result has been rounded to 5 sig digs since the given quantity (953.00 g) has 5 sig digs.

38. a.

$$55 \text{ g HgS} \cdot \frac{\text{mol}}{232.66 \text{ g}} \cdot \frac{6.022 \times 10^{23} \text{ particles}}{\text{mol}} = 1.4 \times 10^{23} \text{ particles}$$

In this ionic compound, a particle is a formula unit. Each formula unit contains one Hg atom, so the number of Hg atoms is equal to the number of particles.

38. b.

$$3.00 \text{ kg Fe}_2\text{O}_3 \cdot \frac{1000 \text{ g}}{1 \text{ kg}} \cdot \frac{\text{mol}}{159.692 \text{ g}} \cdot \frac{6.022 \times 10^{23} \text{ particles}}{\text{mol}} = 1.13 \times 10^{25} \text{ particles}$$

In this ionic compound, a particle is a formula unit. Each formula unit contains two Fe atoms, so the number of Fe atoms is twice the number of particles, or 2.26×10^{25}.

38. c.

$$1.0000 \text{ mol CaCO}_3 \cdot \frac{6.0221 \times 10^{23} \text{ particles}}{\text{mol}} = 6.0221 \times 10^{23} \text{ particles}$$

In this ionic compound, a particle is a formula unit. Each formula unit contains one Ca atom, so the number of Ca atoms is equal to the number of particles. The Avogadro constant was written with 5 sig digs because the given quantity (1.0000 mol) contains 5 sig digs.

38. d

$$45 \text{ g Sr(NO}_2)_2 \cdot \frac{\text{mol}}{179.63 \text{ g}} \cdot \frac{6.022 \times 10^{23} \text{ particles}}{\text{mol}} = 1.5 \times 10^{23} \text{ particles}$$

In this ionic compound, a particle is a formula unit. Each formula unit contains one Sr atom, so the number of Sr atoms is equal to the number of particles.

38. e.

$$2.000 \text{ mol Na}_2\text{CrO}_4 \cdot \frac{6.022 \times 10^{23} \text{ particles}}{\text{mol}} = 1.2044 \times 10^{24} \text{ particles}$$

This result has one extra sig dig. In this ionic compound, a particle is a formula unit. Each formula unit contains two Na atoms, so the number of Na atoms is twice the number of particles, or 2.4088×10^{24}. Rounding to 4 sig digs gives 2.409×10^{24} Na atoms.

38. f.

$$5.05 \text{ kg Ca(CH}_3\text{COO)}_2 \cdot \frac{1000 \text{ g}}{1 \text{ kg}} \cdot \frac{\text{mol}}{158.167 \text{ g}} \cdot \frac{6.022 \times 10^{23} \text{ particles}}{\text{mol}} = 1.92 \times 10^{25} \text{ particles}$$

In this ionic compound, a particle is a formula unit. Each formula unit contains one Ca atom, so the number of Ca atoms is equal to the number of particles.

39.

53.9396 u·0.05845+55.9349 u·0.91754+56.9354 u·0.02119+57.9333 u·0.00282=

$$3.1528\ u+51.323\ u+1.206\ u+0.163\ u=55.845\ u$$

The number of sig digs is driven by the addition rule, the result being limited to three decimal places.

43.

4.62 g+0.776 g+6.154 g=11.55 g (sample mass)

4.62 g/11.55 g=0.400→40.0% C

0.776 g/11.55 g=0.0672→6.72% H

6.154 g/11.55 g=0.5328→53.28% O

The sig digs shown here result from the given quantities. (To add these percentages, decimals are limited by the addition rule to 1 decimal place, so the quantities would have to be rounded to 40.0%, 6.7%, and 53.3%. Arguably, this is also a legitimate way of expressing the result.)

Assuming a 100-gram sample:

$$40.0\ g\ C\cdot\frac{mol}{12.011\ g}=3.33\ mol\ \ (3.33/3.33=1)$$

$$6.72\ g\ H\cdot\frac{mol}{1.0079\ g}=6.67\ mol\ \ (6.67/3.33=2)$$

$$53.28\ g\ O\cdot\frac{mol}{15.9994\ g}=3.330\ mol\ \ (3.330/3.33=1)$$

These ratios give an empirical formula of CH_2O.

1·12.011 u+2·1.0079 u+1·15.9994 u=30.026 u

180.157 u/30.026 u≈6

Applying this multiplier to the empirical formula gives a molecular formula of $C_6H_{12}O_2$.

Chapter 7

16.

$$AgNO_3 + NaBr \rightarrow AgBr + NaNO_3$$

$$3.5 \text{ mol AgNO}_3 \cdot \frac{1 \text{ mol AgBr}}{1 \text{ mol AgNO}_3} = 3.5 \text{ mol AgBr}$$

$$3.5 \text{ mol AgBr} \cdot \frac{187.772 \text{ g}}{\text{mol}} = 657 \text{ g}$$

Rounded to 2 sig digs, the answer is 660 g.

17.

$$2H_2 + O_2 \rightarrow 2H_2O$$

$$1575 \text{ mol H}_2 \cdot \frac{2 \text{ mol H}_2O}{2 \text{ mol H}_2} = 1575 \text{ mol H}_2O$$

$$1575 \text{ mol H}_2 \cdot \frac{1 \text{ mol O}_2}{2 \text{ mol H}_2} = 787.5 \text{ mol O}_2$$

18.

$$Al_2S_3 + 6H_2O \rightarrow 3H_2S + 2Al(OH)_3$$

$$2.290 \text{ kg Al}_2S_3 \cdot \frac{1000 \text{ g}}{1 \text{ kg}} \cdot \frac{\text{mol}}{150.161 \text{ g}} = 15.250 \text{ mol Al}_2S_3$$

$$15.250 \text{ mol Al}_2S_3 \cdot \frac{2 \text{ mol Al(OH)}_3}{1 \text{ mol Al}_2S_3} = 30.500 \text{ mol Al(OH)}_3$$

$$30.50 \text{ mol Al(OH)}_3 \cdot \frac{78.0034 \text{ g}}{\text{mol}} = 2379 \text{ g Al(OH)}_3$$

19.

$$Al(OH)_3 + 3HCl \rightarrow AlCl_3 + 3H_2O$$

19. a.

$$750 \text{ mg Al(OH)}_3 \cdot \frac{1 \text{ g}}{1000 \text{ mg}} \cdot \frac{\text{mol}}{78.0034 \text{ g}} = 0.00961 \text{ mol Al(OH)}_3$$

$$0.00961 \text{ mol Al(OH)}_3 \cdot \frac{3 \text{ mol HCl}}{1 \text{ mol Al(OH)}_3} = 0.0288 \text{ mol HCl}$$

Rounding this result to 2 sig digs gives 0.029 mol HCl.

19. b.

$$750 \text{ mg Al(OH)}_3 \cdot \frac{1 \text{ g}}{1000 \text{ mg}} \cdot \frac{\text{mol}}{78.0034 \text{ g}} = 0.00961 \text{ mol Al(OH)}_3$$

$$0.00961 \text{ mol Al(OH)}_3 \cdot \frac{3 \text{ mol H}_2\text{O}}{1 \text{ mol Al(OH)}_3} = 0.0288 \text{ mol H}_2\text{O}$$

$$0.0288 \text{ mol H}_2\text{O} \cdot \frac{18.02 \text{ g}}{\text{mol}} = 0.5198 \text{ g H}_2\text{O}$$

Rounding this result to 2 sig digs gives 0.52 g H_2O.

20.

$$Fe_2O_3 + 3CO \rightarrow 2Fe + 3CO_2$$

$$2.50 \times 10^4 \text{ kg Fe}_2\text{O}_3 \cdot \frac{1000 \text{ g}}{1 \text{ kg}} \cdot \frac{\text{mol}}{159.692 \text{ g}} = 156,600 \text{ mol Fe}_2\text{O}_3$$

$$156,600 \text{ mol Fe}_2\text{O}_3 \cdot \frac{2 \text{ mol Fe}}{1 \text{ mol Fe}_2\text{O}_3} = 313,100 \text{ mol Fe}$$

$$313,100 \text{ mol Fe} \cdot \frac{55.847 \text{ g}}{\text{mol}} = 1.75 \times 10^7 \text{ g Fe} \cdot \frac{1 \text{ kg}}{1000 \text{ g}} = 1.75 \times 10^4 \text{ kg Fe}$$

21.

$$H_2SO_4 + 2NaOH \rightarrow Na_2SO_4 + 2H_2O$$

21. a.

$$29.55 \text{ g NaOH} \cdot \frac{\text{mol}}{39.9971 \text{ g}} = 0.7388 \text{ mol NaOH}$$

$$44.11 \text{ g H}_2\text{SO}_4 \cdot \frac{\text{mol}}{98.079 \text{ g}} = 0.4497 \text{ mol H}_2\text{SO}_4$$

$$0.4497 \text{ mol H}_2\text{SO}_4 \cdot \frac{2 \text{ mol NaOH}}{1 \text{ mol H}_2\text{SO}_4} = 0.8994 \text{ mol NaOH required}$$

This much NaOH is not available, so NaOH is the limiting reactant.

---

21. b.

$$0.73880 \text{ mol NaOH} \cdot \frac{1 \text{ mol Na}_2\text{SO}_4}{2 \text{ mol NaOH}} = 0.36940 \text{ mol Na}_2\text{SO}_4$$

$$0.36940 \text{ mol Na}_2\text{SO}_4 \cdot \frac{142.043 \text{ g}}{\text{mol}} = 52.47 \text{ g Na}_2\text{SO}_4$$

---

21. c.

$$0.73880 \text{ mol NaOH} \cdot \frac{2 \text{ mol H}_2\text{O}}{2 \text{ mol NaOH}} = 0.73880 \text{ mol H}_2\text{O}$$

$$0.73880 \text{ mol H}_2\text{O} \cdot \frac{18.015 \text{ g}}{\text{mol}} = 13.31 \text{ g H}_2\text{O}$$

---

22. a.

$$350.0 \text{ mol } C_8H_{18} \cdot \frac{25 \text{ mol } O_2}{2 \text{ mol } C_8H_{18}} = 4375 \text{ mol } O_2$$

The result has been rounded to 4 sig digs as required by the given information.

22. b.

$$3.5 \text{ kg } O_2 \cdot \frac{1000 \text{ g}}{1 \text{ kg}} \cdot \frac{\text{mol}}{32.00 \text{ g}} = 109.4 \text{ mol } O_2$$

$$109.4 \text{ mol } O_2 \cdot \frac{18 \text{ mol } H_2O}{25 \text{ mol } O_2} = 78.77 \text{ mol } H_2O$$

$$78.77 \text{ mol } H_2O \cdot \frac{18.02 \text{ g}}{\text{mol}} \cdot \frac{1 \text{ kg}}{1000 \text{ g}} = 1.4 \text{ kg } H_2O$$

The result has been rounded to 2 sig digs as required by the given information.

22. c.

$$11.5 \text{ gal } C_8H_{18} \cdot \frac{3.785 \text{ L}}{1 \text{ gal}} \cdot \frac{1000 \text{ mL}}{1 \text{ L}} \cdot \frac{0.692 \text{ g}}{\text{mL}} = 30,120 \text{ g } C_8H_{18}$$

$$30,120 \text{ g } C_8H_{18} \cdot \frac{\text{mol}}{114.230 \text{ g}} = 263.7 \text{ mol } C_8H_{18}$$

$$263.7 \text{ mol } C_8H_{18} \cdot \frac{25 \text{ mol } O_2}{2 \text{ mol } C_8H_{18}} = 3296 \text{ mol } O_2$$

$$3296 \text{ mol } O_2 \cdot \frac{32.00 \text{ g}}{\text{mol}} \cdot \frac{1 \text{ kg}}{1000 \text{ g}} = 105 \text{ kg } O_2$$

The result has been rounded to 3 sig digs as required by the given information.

23. a.

$$855 \text{ g } NH_3 \cdot \frac{\text{mol}}{17.0304 \text{ g}} = 50.20 \text{ mol } NH_3$$

$$1750 \text{ g } O_2 \cdot \frac{\text{mol}}{32.00 \text{ g}} = 54.69 \text{ mol } O_2$$

$$50.20 \text{ mol } NH_3 \cdot \frac{5 \text{ mol } O_2}{4 \text{ mol } NH_3} = 62.75 \text{ mol } O_2 \text{ required.}$$

This much O_2 is not available, so O_2 is the limiting reactant.

23. b.

$$1750 \text{ g O}_2 \cdot \frac{\text{mol}}{32.00 \text{ g}} = 54.69 \text{ mol O}_2$$

$$54.69 \text{ mol O}_2 \cdot \frac{4 \text{ mol NO}}{5 \text{ mol O}_2} = 43.75 \text{ mol NO}$$

$$43.75 \text{ mol NO} \cdot \frac{30.006 \text{ g}}{\text{mol}} = 1313 \text{ g NO}$$

Rounding to 3 sig digs as required by the given information, we have 1,310 g NO.

23. c.

$$\frac{1272 \text{ g}}{1313 \text{ g}} \cdot 100\% = 96.9\%$$

24.

$$2\text{Al(OH)}_3 + 3\text{H}_2\text{SO}_4 \rightarrow \text{Al}_2(\text{SO}_4)_3 + 6\text{H}_2\text{O}$$

24. a.

$$31.8 \text{ g H}_2\text{SO}_4 \cdot \frac{\text{mol}}{98.079 \text{ g}} = 0.3242 \text{ mol H}_2\text{SO}_4$$

$$25.4 \text{ g Al(OH)}_3 \cdot \frac{\text{mol}}{78.0034 \text{ g}} = 0.3256 \text{ mol Al(OH)}_3$$

$$0.3256 \text{ mol Al(OH)}_3 \cdot \frac{3 \text{ mol H}_2\text{SO}_4}{2 \text{ mol Al(OH)}_3} = 0.4884 \text{ mol H}_2\text{SO}_4 \text{ required.}$$

This much H_2SO_4 is not available, so H_2SO_4 is the limiting reactant.

24. b.

$$0.3242 \text{ mol H}_2\text{SO}_4 \cdot \frac{2 \text{ mol Al(OH)}_3}{3 \text{ mol H}_2\text{SO}_4} = 0.2161 \text{ mol Al(OH)}_3$$

$$0.2161 \text{ mol Al(OH)}_3 \cdot \frac{78.0034 \text{ g}}{\text{mol}} = 16.86 \text{ g Al(OH)}_3 \text{ are consumed}$$

$25.4 \text{ g} - 16.9 \text{ g} = 8.5 \text{ g Al(OH)}_3$ remaining

24. c.

$$0.3242 \text{ mol H}_2\text{SO}_4 \cdot \frac{1 \text{ mol Al}_2(\text{SO}_4)_3}{3 \text{ mol H}_2\text{SO}_4} = 0.1081 \text{ mol Al}_2(\text{SO}_4)_3$$

$$0.1081 \text{ mol Al}_2(\text{SO}_4)_3 \cdot \frac{342.154 \text{ g}}{\text{mol}} = 37.0 \text{ g Al}_2(\text{SO}_4)_3 \text{ produced}$$

$$0.3242 \text{ mol H}_2\text{SO}_4 \cdot \frac{6 \text{ mol H}_2\text{O}}{3 \text{ mol H}_2\text{SO}_4} = 0.6484 \text{ mol H}_2\text{O}$$

$$0.6484 \text{ mol H}_2\text{O} \cdot \frac{18.02 \text{ g}}{\text{mol}} = 11.7 \text{ g H}_2\text{O produced}$$

25. a.

$$1.200\times10^3 \text{ kg N}_2\text{H}_4 \cdot \frac{1000\text{ g}}{1\text{ kg}} \cdot \frac{\text{mol}}{32.0368\text{ g N}_2\text{H}_4} = 37{,}457 \text{ mol N}_2\text{H}_4$$

$$1.000\times10^3 \text{ kg (CH}_3)_2\text{N}_2\text{H}_2 \cdot \frac{1000\text{ g}}{1\text{ kg}} \cdot \frac{\text{mol}}{60.099\text{ g N}_2\text{H}_4} = 16{,}639 \text{ mol (CH}_3)_2\text{N}_2\text{H}_2$$

$$4.500\times10^3 \text{ kg N}_2\text{O}_4 \cdot \frac{1000\text{ g}}{1\text{ kg}} \cdot \frac{\text{mol}}{92.0110\text{ g N}_2\text{O}_4} = 48{,}907 \text{ mol N}_2\text{O}_4$$

$$37{,}457 \text{ mol N}_2\text{H}_4 \cdot \frac{1 \text{ mol (CH}_3)_2\text{N}_2\text{H}_2}{2 \text{ mol N}_2\text{H}_4} = 18{,}729 \text{ mol (CH}_3)_2\text{N}_2\text{H}_2$$

This much $(\text{CH}_3)_2\text{N}_2\text{H}_2$ is not available, so $(\text{CH}_3)_2\text{N}_2\text{H}_2$ is the limiting reactant between these two reactants. Next, we compare this with N_2O_4:

$$16{,}639 \text{ mol (CH}_3)_2\text{N}_2\text{H}_2 \cdot \frac{3 \text{ mol N}_2\text{O}_4}{1 \text{ mol (CH}_3)_2\text{N}_2\text{H}_2} = 49{,}917 \text{ mol N}_2\text{O}_4$$

This much N_2O_4 is not available, so N_2O_4 is the limiting reactant for the reaction, and was thus consumed first.

25. b.

$$48{,}907 \text{ mol N}_2\text{O}_4 \cdot \frac{8 \text{ mol H}_2\text{O}}{3 \text{ mol N}_2\text{O}_4} = 130{,}491 \text{ mol H}_2\text{O}$$

$$130{,}491 \text{ mol H}_2\text{O} \cdot \frac{18.015\text{ g}}{\text{mol}} \cdot \frac{1\text{ kg}}{1000\text{ g}} = 2349 \text{ kg H}_2\text{O}$$

25. c.

$$48{,}907 \text{ mol N}_2\text{O}_4 \cdot \frac{6 \text{ mol N}_2}{3 \text{ mol N}_2\text{O}_4} = 97{,}814 \text{ mol N}_2$$

$$97{,}814 \text{ mol N}_2 \cdot \frac{6.0221\times10^{23} \text{ particles}}{\text{mol}} = 5.890\times10^{28} \text{ particles N}_2$$

Since there are 2 N atoms per molecule, this gives 1.178×10^{29} N atoms.

26. a.

$$542 \text{ g SiO}_2 \cdot \frac{\text{mol}}{60.0843\text{ g}} = 9.021 \text{ mol}$$

$$9.021 \text{ mol SiO}_2 \cdot \frac{6 \text{ mol HF}}{1 \text{ mol SiO}_2} = 54.1 \text{ mol HF}$$

26. b.

$$4.25 \text{ mol HF} \cdot \frac{1 \text{ mol H}_2\text{SiF}_6}{6 \text{ mol HF}} = 0.7083 \text{ mol H}_2\text{SiF}_6$$

$$0.7083 \text{ mol H}_2\text{SiF}_6 \cdot \frac{144.0917\text{ g}}{\text{mol}} = 102 \text{ g H}_2\text{SiF}_6$$

26. c.

$$2.0 \text{ gal} \cdot \frac{3.785 \text{ L}}{1 \text{ gal}} \cdot \frac{1000 \text{ mL}}{1 \text{ L}} \cdot \frac{0.998 \text{ g}}{\text{mL}} = 7555 \text{ g H}_2\text{O}$$

$$7555 \text{ g H}_2\text{O} \cdot \frac{\text{mol}}{18.015 \text{ g}} = 419 \text{ mol H}_2\text{O}$$

$$419 \text{ mol H}_2\text{O} \cdot \frac{6 \text{ mol HF}}{2 \text{ mol H}_2\text{O}} = 1300 \text{ mol HF}$$

The result has been rounded to 2 sig digs as required.

26. d.

$$2.50 \text{ kg HF} \cdot \frac{1000 \text{ g}}{1 \text{ kg}} \cdot \frac{\text{mol}}{20.0063 \text{ g}} = 125.0 \text{ mol HF}$$

$$1.205 \text{ kg SiO}_2 \cdot \frac{1000 \text{ g}}{1 \text{ kg}} \cdot \frac{\text{mol}}{60.0843 \text{ g}} = 20.06 \text{ mol SiO}_2$$

$$20.06 \text{ mol SiO}_2 \cdot \frac{6 \text{ mol HF}}{1 \text{ mol SiO}_2} = 120.3 \text{ mol HF}$$

More HF than this is available, so the limiting reactant is SiO_2.

26. e.

$$20.06 \text{ mol SiO}_2 \cdot \frac{1 \text{ mol H}_2\text{SiF}_6}{1 \text{ mol SiO}_2} = 20.06 \text{ mol H}_2\text{SiF}_6$$

$$20.06 \text{ mol H}_2\text{SiF}_6 \cdot \frac{144.0917 \text{ g}}{\text{mol}} = 2890 \text{ g} \cdot \frac{1 \text{ kg}}{1000 \text{ g}} = 2.89 \text{ kg H}_2\text{SiF}_6$$

26. f.

$$\frac{2.657 \text{ kg}}{2.890 \text{ kg}} \cdot 100\% = 91.9\%$$

27.

$$\text{C}_6\text{H}_6 + \text{Br}_2 \rightarrow \text{C}_6\text{H}_5\text{Br} + \text{HBr}$$

27. a.

$$45.0 \text{ g C}_6\text{H}_6 \cdot \frac{\text{mol}}{78.113 \text{ g}} = 0.5761 \text{ mol C}_6\text{H}_6$$

$$97.5 \text{ g Br}_2 \cdot \frac{\text{mol}}{159.808 \text{ g}} = 0.6101 \text{ mol Br}_2$$

$$0.5761 \text{ mol C}_6\text{H}_6 \cdot \frac{1 \text{ mol Br}_2}{1 \text{ mol C}_6\text{H}_6} = 0.5761 \text{ mol Br}_2$$

More Br_2 than this is available, so C_6H_6 is the limiting reactant.

$$0.5761 \text{ mol } C_6H_6 \cdot \frac{1 \text{ mol } C_6H_5Br}{1 \text{ mol } C_6H_6} = 0.5761 \text{ mol } C_6H_5Br$$

$$0.5761 \text{ mol } C_6H_5Br \cdot \frac{157.010 \text{ g}}{\text{mol}} = 90.5 \text{ g } C_6H_5Br$$

27. b.

$$\frac{63.25 \text{ g}}{90.5 \text{ g}} \cdot 100\% = 69.9\%$$

28. a.

$$2.85 \times 10^{-6} \text{ mol } O_3 \cdot \frac{2 \text{ mol NaI}}{1 \text{ mol } O_3} = 5.70 \times 10^{-6} \text{ mol NaI}$$

28. b.

$$3.00 \text{ g } O_3 \cdot \frac{\text{mol}}{48.00 \text{ g}} = 0.06250 \text{ mol } O_3$$

$$0.06250 \text{ mol } O_3 \cdot \frac{2 \text{ mol NaI}}{1 \text{ mol } O_3} = 0.1250 \text{ mol NaI}$$

$$0.1250 \text{ mol NaI} \cdot \frac{149.08 \text{ g}}{\text{mol}} = 18.6 \text{ g NaI}$$

28. c.

$$1455 \text{ g NaI} \cdot \frac{\text{mol}}{149.8943 \text{ g}} = 9.7068 \text{ mol NaI}$$

$$250.0 \text{ g } O_3 \cdot \frac{\text{mol}}{47.998 \text{ g}} = 5.2085 \text{ mol } O_3$$

$$5.2086 \text{ mol } O_3 \cdot \frac{2 \text{ mol NaI}}{1 \text{ mol } O_3} = 10.417 \text{ mol NaI}$$

This much NaI is not available, so NaI is the limiting reactant.

$$9.7068 \text{ mol NaI} \cdot \frac{1 \text{ mol } I_2}{2 \text{ mol NaI}} = 4.8534 \text{ mol } I_2$$

$$4.8534 \text{ mol } I_2 \cdot \frac{253.809 \text{ g}}{\text{mol}} = 1232 \text{ g } I_2$$

28. d.

$$4.8534 \text{ mol } I_2 \cdot \frac{6.0221 \times 10^{23} \text{ particles}}{\text{mol}} = 2.923 \times 10^{24} \text{ particles}$$

Each particle is a molecules containing 2 I atoms, so there are 5.846 × 10²⁴ I atoms.

29. a.

16.81 mg / 25.0 mg = 0.672 → 67.2% C

1.736 mg / 25.0 g = 0.0694 → 6.94% H

3.015 mg / 25.0 g = 0.121 → 12.1% N

3.445 mg / 25.0 g = 0.138 → 13.8% O

---

29. b.

Assuming a 100-gram sample:

$$67.2 \text{ g C} \cdot \frac{\text{mol}}{12.011 \text{ g}} = 5.59 \text{ mol} \quad (5.59/0.863 \approx 6.5) \rightarrow 6.5 \cdot 2 = 13$$

$$6.94 \text{ g H} \cdot \frac{\text{mol}}{1.0079 \text{ g}} = 6.89 \text{ mol} \quad (6.89/0.863 \approx 8) \rightarrow 8 \cdot 2 = 16$$

$$12.1 \text{ g N} \cdot \frac{\text{mol}}{14.0067 \text{ g}} = 0.864 \text{ mol} \quad (0.864/0.863 \approx 1) \rightarrow 1 \cdot 2 = 2$$

$$13.8 \text{ g O} \cdot \frac{\text{mol}}{15.9994 \text{ g}} = 0.863 \text{ mol} \quad (0.863/0.863 = 1) \rightarrow 1 \cdot 2 = 2$$

Proportions are doubled to give whole numbers. These ratios give an empirical formula of $C_{13}H_{16}N_2O_2$. The formula mass of this formula is

$$13 \cdot 12.011 \text{ u} + 16 \cdot 1.0079 \text{ u} + 2 \cdot 14.0067 \text{ u} + 2 \cdot 15.9994 = 232.282 \text{ u}$$

This is the same as the given molar mass. Thus, the molecular formula is $C_{13}H_{16}N_2O_2$.

---

29. c.

$$3.0 \text{ mg } C_{13}H_{16}N_2O_2 \cdot \frac{1 \text{ g}}{1000 \text{ mg}} \cdot \frac{\text{mol}}{232 \text{ g}} \cdot \frac{6.022 \times 10^{23} \text{ particles}}{\text{mol}} = 7.79 \times 10^{18} \text{ particles}$$

Each particle (molecule) contains 13 C atoms, giving 1.0×10^{20} C atoms.

30.

$$E = 2.271 \text{ eV} \cdot \frac{1.60218 \times 10^{-19} \text{ J}}{1 \text{ eV}} = 3.6386 \times 10^{-19} \text{ J}$$

$$E = \frac{hv}{\lambda} \rightarrow \lambda = \frac{hv}{E} = \frac{6.626 \times 10^{-34} \text{ J} \cdot \text{s} \cdot 2.9979 \times 10^8 \ \frac{\text{m}}{\text{s}}}{3.6386 \times 10^{-19} \text{ J}} = 5.459 \times 10^{-7} \text{ m} \cdot \frac{1 \times 10^9 \text{ nm}}{1 \text{ m}} = 545.9 \text{ nm}$$

33.

$$27.9769 \text{ u} \cdot 0.92223 + 28.9765 \text{ u} \cdot 0.04685 + 29.9738 \text{ u} \cdot 0.03092 =$$

$$25.801 \text{ u} + 1.358 \text{ u} + 0.9268 \text{ u} = 28.085 \text{ u}$$

Chapter 8

5.

Since we are using the density in a calculation with the acceleration of gravity and pressure, we must use MKS units.

$$\rho = 0.998 \ \frac{g}{mL} \cdot \frac{1 \ kg}{1000 \ g} \cdot \frac{1000 \ mL}{1 \ L} \cdot \frac{1000 \ L}{m^3} = 998 \ \frac{kg}{m^3}$$

$$g = 9.80 \ \frac{m}{s^2}$$

$$P = 101,325 \ Pa$$

$$P = \rho g h \rightarrow h = \frac{P}{\rho g} = \frac{101,325 \ Pa}{998 \ \frac{kg}{m^3} \cdot 9.80 \ \frac{m}{s^2}} = 10.4 \ m$$

6.

$$1850 \ psi \cdot \frac{6894.8 \ Pa}{1 \ psi} = 12,755,000 \ Pa$$

Rounding to 3 sig digs, we have 12,800,000 Pa.

$$12,800,000 \ Pa \cdot \frac{1 \ kPa}{1000 \ Pa} = 12,800 \ kPa$$

$$12,755,000 \ Pa \cdot \frac{1 \ atm}{101,325 \ Pa} = 125.89 \ atm$$

Rounding to 3 sig digs, we have 126 atm.

$$125.89 \ atm \cdot \frac{760 \ Torr}{1 \ atm} = 95,700 \ Torr$$

$$12,800,000 \ Pa \cdot \frac{1 \ bar}{100,000 \ Pa} = 128 \ bar$$

7.

Using the symbol F_w to represent the skater's weight:

$$l = 2.9 \ mm \cdot \frac{1 \ m}{1000 \ mm} = 0.0029 \ m$$

$$w = 57 \ mm \cdot \frac{1 \ m}{1000 \ mm} = 0.057 \ m$$

$$F_w = 200.0 \ lb \cdot \frac{4.448 \ N}{1 \ lb} = 889.6 \ N$$

$$A = l \cdot w = 0.0029 \ m \cdot 0.0057 \ m = 0.0001653 \ m^2$$

$$P = \frac{F}{A} = \frac{889.6 \ N}{0.0001653 \ m^2} = 5,382,000 \ Pa$$

$$5,382,000 \ Pa \cdot \frac{1 \ atm}{101,325 \ Pa} = 53 \ atm$$

19. a.

$n = 3.47$ mol

$\Delta T = T_f - T_i = 55°C - 25°C = 30°C = 30$ K (with 2 sig digs)

$C = 0.0752 \dfrac{kJ}{mol \cdot K}$

$Q = Cn\Delta T = 0.0752 \dfrac{kJ}{mol \cdot K} \cdot 3.47 \text{ mol} \cdot 30 \text{ K} = 7.8$ kJ

This value has been rounded to 2 sig digs as the given temperature data require.

---

19. b.

$Q = 3.65 \times 10^5$ kJ

$H_v = 40.7 \dfrac{kJ}{mol}$

$Q = mH_v \rightarrow m = \dfrac{Q}{H_v} = \dfrac{3.65 \times 10^5 \text{ kJ}}{40.7 \dfrac{kJ}{mol}} = 8970$ mol

$8970 \text{ mol} \cdot \dfrac{18.02 \text{ g}}{mol} \cdot \dfrac{1 \text{ kg}}{1000 \text{ g}} = 162$ kg

---

19. c.

$m = 12.00 \text{ kg} \cdot \dfrac{1000 \text{ g}}{1 \text{ kg}} = 12,000$ g (with 4 sig digs)

$H_f = 6.01 \dfrac{kJ}{mol}$

$n = 12,000 \text{ g} \cdot \dfrac{mol}{18.015 \text{ g}} = 666.11$ mol

$Q = nH_f = 666.11 \text{ mol} \cdot 6.01 \dfrac{kJ}{mol} = 4003$ kJ

To round this value to 3 sig digs, we must write it in scientific notation, giving 4.00×10^3 kJ.

---

19. d.

$Q = 7.32$ kJ

$\Delta T = T_f - T_i = 30.0°C - 20.0°C = 10.0°C = 10.0$ K

$C = 0.0752 \ \dfrac{\text{kJ}}{\text{mol} \cdot \text{K}}$

$\rho = 0.9970 \ \dfrac{\text{g}}{\text{cm}^3}$

$Q = Cn\Delta T \rightarrow n = \dfrac{Q}{C\Delta T} = \dfrac{7.32 \text{ kJ}}{0.0752 \ \dfrac{\text{kJ}}{\text{mol} \cdot \text{K}} \cdot 10.0 \text{ K}} = 9.734$ mol

$9.734 \text{ mol} \cdot \dfrac{18.015 \text{ g}}{\text{mol}} = 175.4$ g

$\rho = \dfrac{m}{V} \rightarrow V = \dfrac{m}{\rho} = \dfrac{175.4 \text{ g}}{0.9970 \ \dfrac{\text{g}}{\text{cm}^3}} = 176 \text{ cm}^3 = 176$ mL

---

19. e.

$m = 5.00$ kg

$\Delta T_{ice} = T_f - T_i = 0.0°C - (-25.0°C) = 25.0°C = 25.0$ K

$C_{ice} = 0.0364 \ \dfrac{\text{kJ}}{\text{mol} \cdot \text{K}}$

$H_f = 6.01 \ \dfrac{\text{kJ}}{\text{mol}}$

Begin by computing the number of moles:

$5.00 \text{ kg H}_2\text{O} \cdot \dfrac{1000 \text{ g}}{1 \text{ kg}} \cdot \dfrac{\text{mol}}{18.015 \text{ g}} = 277.5$ mol

Next compute the heat required to raise the temperature of the ice from –25°C to 0°C:

$Q = Cn\Delta T = 0.0364 \ \dfrac{\text{kJ}}{\text{mol} \cdot \text{K}} \cdot 277.5 \text{ mol} \cdot 25.0 \text{ K} = 252.5$ kJ

This value has one extra sig digs which we will round off later. Next compute the heat required to melt the ice:

$Q = nH_f = 277.5 \text{ mol} \cdot 6.10 \ \dfrac{\text{kJ}}{\text{mol}} = 1668$ kJ

This value also has one extra sig dig. The answer comes when we add these heat values (using the addition rule for the sig digs). To do this, we need to first round off each value to the correct number of sig digs, then add them using the rule.

1670 kJ

+ 253 kJ

1920 kJ

---

20. a.

The dotted line at atmospheric pressure (1 atm) indicates sublimation occurs at approximately −75°C.

20. b.

Using the sample scale as a reference, we estimate the triple point pressure to be 4 atm. Converting to Torr,

$$4 \text{ atm} \cdot \frac{760 \text{ Torr}}{1 \text{ atm}} = 3000 \text{ Torr}$$

Rounded to 1 sig dig this would be 400 Torr. The triple-point temperature is approximately −55°C, which is 218 K.

20. c.

$$7600 \text{ Torr} \cdot \frac{1 \text{ atm}}{760 \text{ Torr}} = 10 \text{ atm}$$

At this pressure, the temperatures at the solid-liquid and liquid-gas boundaries are approximately −55°C and −35°C, respectively.

20. d.

This is the pressure at the tip of the liquid-gas curve, approximately 40 atm.

20. e.

Again using the logarithmic scale for reference, a vertical line at −20°C crosses the gas-liquid and liquid-solid curves at approximately 15 atm and 1000 atm, respectively.

20. f.

The vapor pressure is the pressure at the liquid-gas curve at a given temperature, or approximately 35 atm in this case.

20. g.

That pressure is above the solid-liquid curve, so the gas would solidify.

24. a.

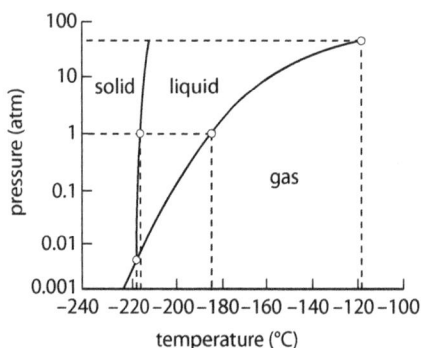

24. b.

The slope of the solid-liquid curve from the triple point to atmospheric pressure is positive, not negative as it is with water. This implies that solid oxygen is denser than liquid oxygen and will not float the way water ice does.

24. c.

At atmospheric pressure, heating moves the conditions of solid oxygen across the solid-liquid curve, and thus heating solid O_2 causes it to melt.

25.

The given pressure of 594 Torr falls between the pressures of 567.7 Torr and 611.6 Torr shown in Table A.4. Thus, the boiling point falls between the temperatures associated with these vapor pressures (92°C and 94°C), putting the boiling point of water at approximately 93°C.

27. a.

$$5.00 \text{ mol } C_8H_{18} \cdot \frac{25 \text{ mol } O_2}{2 \text{ mol } C_8H_{18}} = 62.5 \text{ mol } O_2$$

27. b.

$$\rho = 0.692 \frac{g}{mL}$$

$$1.00 \text{ gal } C_8H_{18} \cdot \frac{3.785 \text{ L}}{gal} \cdot \frac{1000 \text{ mL}}{L} = 3785 \text{ mL}$$

$$\rho = \frac{m}{V} \rightarrow m = \rho V = 0.692 \frac{g}{mL} \cdot 3785 \text{ mL} = 2619 \text{ g}$$

$$2619 \text{ g } C_8H_{18} \cdot \frac{mol}{114.23 \text{ g}} = 22.93 \text{ mol}$$

$$22.93 \text{ mol } C_8H_{18} \cdot \frac{25 \text{ mol } O_2}{2 \text{ mol } C_8H_{18}} = 286.6 \text{ mol } O_2$$

$$286.6 \text{ mol } O_2 \cdot \frac{32.00 \text{ g}}{mol} = 9170 \text{ g}$$

An extra sig dig was used during the calculation, and the result was rounded to 3 sig digs as the given information requires.

29. a.

$$2.50 \text{ g } Zn \cdot \frac{mol}{65.39 \text{ g}} = 0.03823 \text{ mol } Zn$$

$$3.00 \text{ g } AgNO_3 \cdot \frac{mol}{169.9 \text{ g}} = 0.01766 \text{ mol } AgNO_3$$

$$0.03823 \text{ mol } Zn \cdot \frac{2 \text{ mol } AgNO_3}{1 \text{ mol } Zn} = 0.07646 \text{ mol } AgNO_3$$

This much $AgNO_3$ is not available, so $AgNO_3$ is the limiting reactant.

29. b.

$$0.01766 \text{ mol } AgNO_3 \cdot \frac{1 \text{ mol } Zn(NO_3)_2}{2 \text{ mol } AgNO_3} = 0.00883 \text{ mol } Zn(NO_3)_2$$

$$0.00883 \text{ mol } Zn(NO_3)_2 \cdot \frac{189.40 \text{ g}}{mol} = 1.67 \text{ g}$$

29. c.

$$0.01766 \text{ mol AgNO}_3 \cdot \frac{1 \text{ mol Zn}}{2 \text{ mol AgNO}_3} = 0.00883 \text{ mol Zn}$$

$$0.03823 \text{ mol} - 0.00883 \text{ mol} = 0.0294 \text{ mol Zn remaining}$$

$$0.0294 \text{ mol Zn} \cdot \frac{65.39 \text{ g}}{\text{mol}} = 1.92 \text{ g}$$

30.

Assuming a 100-gram sample:

$$75.69 \text{ g C} \cdot \frac{\text{mol}}{12.011 \text{ g}} = 6.302 \text{ mol} \quad (6.302/0.969 \approx 6.5) \rightarrow 6.5 \cdot 2 = 13$$

$$8.80 \text{ g H} \cdot \frac{\text{mol}}{1.0079 \text{ g}} = 8.731 \text{ mol} \quad (8.731/0.969 \approx 9) \rightarrow 9 \cdot 2 = 18$$

$$15.51 \text{ g O} \cdot \frac{\text{mol}}{15.9994 \text{ g}} = 0.969 \text{ mol} \quad (0.969/0.969 = 1) \rightarrow 1 \cdot 2 = 2$$

Proportions are doubled to give whole numbers. These ratios give an empirical formula of $C_{13}H_{18}O_2$. The formula mass of this formula is 206.284 u, the same as the given molar mass. Thus, this is the molecular formula.

Chapter 9

1. a.

$V = 2.00$ L

$T = 22.0°C$

$P_1 = 0.974$ atm

$P_2 = 0.772$ atm

$$P_1V_1 = P_2V_2 \rightarrow V_2 = V_1 \cdot \frac{P_1}{P_2} = 2.00 \text{ L} \cdot \frac{0.974 \text{ atm}}{0.772 \text{ atm}} = 2.52 \text{ L}$$

---

1. b.

$V_1 = 2.00$ L

$V_2 = 3.14$ L

$T_1 = 22.0°C \rightarrow 22.0 + 273.2 = 295.2$ K

$P = 0.974$ atm

$$\frac{V_1}{T_1} = \frac{V_2}{T_2} \rightarrow T_2 = T_1 \cdot \frac{V_2}{V_1} = 295.2 \text{ K} \cdot \frac{3.14 \text{ L}}{2.00 \text{ L}} = 463 \text{ K}$$

2.

$V_1 = 3.5 \text{ m}^3$

$V_2 = 1.25 \text{ m}^3$

$P_1 = 0.275$ bar

$$P_1V_1 = P_2V_2 \rightarrow P_2 = P_1 \cdot \frac{V_1}{V_2} = 0.275 \text{ bar} \cdot \frac{3.5 \text{ m}^3}{1.25 \text{ m}^3} = 0.77 \text{ bar}$$

3.

Stage 1: Volume at the end of this stage is V_2.

$V_1 = 2225 \text{ cm}^3$

$T_1 = 37.00°C \rightarrow 37.00 + 273.15 = 310.15$ K

$T_2 = -15°C \rightarrow -15 + 273.2 = 258.2$ K

$$\frac{V_1}{T_1} = \frac{V_2}{T_2} \rightarrow V_2 = V_1 \cdot \frac{T_2}{T_1} = 2225 \text{ cm}^3 \cdot \frac{258.2 \text{ K}}{310.15 \text{ K}} = 1852 \text{ cm}^3$$

Stage 2: Volumes at beginning and end of this stage are V_2 and V_3.

$P_2 = 655$ Torr

$P_3 = 603$ Torr

$V_2 = 1852 \text{ cm}^3$

$$P_2V_2 = P_3V_3 \rightarrow V_3 = V_2 \cdot \frac{P_2}{P_3} = 1852 \text{ cm}^3 \cdot \frac{655 \text{ Torr}}{603 \text{ Torr}} = 2010 \text{ cm}^3$$

9.

$n = 100.0$ mol

$T = 0°C = 273.15$ K

$P = 100$ kPa

$R = 8.314 \dfrac{\text{L·kPa}}{\text{mol·K}}$

$PV = nRT \rightarrow V = \dfrac{nRT}{P} = \dfrac{100.0 \text{ mol·} 8.314 \dfrac{\text{L·kPa}}{\text{mol·K}} \cdot 273.15 \text{ K}}{100 \text{ kPa}} = 2271 \text{ L}$

10.

$n = 0.105$ mol

$V = 1750 \text{ mL·} \dfrac{1 \text{ L}}{1000 \text{ mL}} = 1.75 \text{ L}$

$P = 15.0 \text{ bar·} \dfrac{100 \text{ kPa}}{1 \text{ bar}} = 1500 \text{ kPa}$

$R = 8.314 \dfrac{\text{L·kPa}}{\text{mol·K}}$

$PV = nRT \rightarrow T = \dfrac{PV}{nR} = \dfrac{1500 \text{ kPa·} 1.75 \text{ L}}{0.105 \text{ mol·} 8.314 \dfrac{\text{L·kPa}}{\text{mol·K}}} = 3007 \text{ K}$

With 3 sig digs this is 3010 K, with the most precise digit in the 10s column.

After the additions below, we will need to round to the 10s column.

$T_C = T_K - 273 = 3007 - 273 = 2734°C$

$T_F = \dfrac{9}{5}T_C + 32 = \dfrac{9}{5} \cdot 2734 + 32 = 4953°F$

Rounding to the 10s column, we have 4,950°F.

11.

$m = 2.10$ g

$P = 145$ Torr

$T = 276$ K

$R = 62.36 \dfrac{\text{L·Torr}}{\text{mol·K}}$

$2.10 \text{ g CH}_4 \cdot \dfrac{\text{mol}}{16.043 \text{ g}} = 0.1309 \text{ mol} = n$

$PV = nRT \rightarrow V = \dfrac{nRT}{P} = \dfrac{0.1309 \text{ mol·} 62.36 \dfrac{\text{L·Torr}}{\text{mol·K}} \cdot 276 \text{ K}}{145 \text{ Torr}} = 15.5 \text{ L}$

12.

$V = 1.75$ L

$P = 0.922 \text{ bar} \cdot \dfrac{100 \text{ kPa}}{1 \text{ bar}} = 922 \text{ kPa}$

$T = 5°C \rightarrow 5°C + 273.15 = 278.2$ K

$R = 8.314 \dfrac{\text{L} \cdot \text{kPa}}{\text{mol} \cdot \text{K}}$

$PV = nRT \rightarrow n = \dfrac{PV}{RT} = \dfrac{922 \text{ kPa} \cdot 1.75 \text{ L}}{8.314 \dfrac{\text{L} \cdot \text{kPa}}{\text{mol} \cdot \text{K}} \cdot 278.2 \text{ K}} = 0.698 \text{ mol}$

---

13. a.

$V = 1550 \text{ m}^3$

$P = 100,000$ Pa

$T = 0°C \rightarrow 0°C + 273.15 = 273.15$ K

$R = 8.314 \dfrac{\text{J}}{\text{mol} \cdot \text{K}}$

$PV = nRT \rightarrow n = \dfrac{PV}{RT} = \dfrac{100,000 \text{ Pa} \cdot 1550 \text{ m}^3}{8.314 \dfrac{\text{J}}{\text{mol} \cdot \text{K}} \cdot 273.15 \text{ K}} = 68,300 \text{ mol}$

---

13. b.

$68,300 \text{ mol CH}_4 \cdot \dfrac{16.043 \text{ g}}{\text{mol}} = 1,095,737 \text{ g} \cdot \dfrac{1 \text{ kg}}{1000 \text{ g}} = 1096 \text{ kg}$

When rounded to 3 sig digs, this gives $\boxed{1.10 \times 10^3 \text{ kg}}$

---

13. c.

$\rho = \dfrac{m}{V} = \dfrac{1096 \text{ kg}}{1550 \text{ m}^3} = 0.707 \dfrac{\text{kg}}{\text{m}^3}$

---

13. d.

$68,300 \text{ mol} \cdot \dfrac{6.022 \times 10^{23} \text{ particles}}{\text{mol}} = 4.11 \times 10^{28} \text{ particles}$

There is one carbon atom in each molecule of CH_4, giving a total of 4.11×10^{28} C atoms.

---

14. a.

$$V = 4960 \text{ m}^3 \cdot \frac{1000 \text{ L}}{1 \text{ m}^3} = 4{,}960{,}000 \text{ L}$$

$$T = 17°C \rightarrow 17°C + 273.15 = 290.2 \text{ K}$$

$$P = 1.00 \text{ atm}$$

$$R = 0.08206 \frac{\text{L·atm}}{\text{mol·K}}$$

$$PV = nRT \rightarrow n = \frac{PV}{RT} = \frac{1.00 \text{ atm} \cdot 4{,}960{,}000 \text{ L}}{0.08206 \dfrac{\text{L·atm}}{\text{mol·K}} \cdot 290.2 \text{ K}} = 208{,}300 \text{ mol}$$

$$208{,}300 \text{ mol He} \cdot \frac{4.003 \text{ g}}{\text{mol}} = 834{,}000 \text{ g} \cdot \frac{1 \text{ kg}}{1000 \text{ g}} = 834 \text{ kg}$$

14. b.

The number of moles would be the same. Hydrogen gas is diatomic, so the mass for H_2 is 2.0158 g/mol.

$$208{,}300 \text{ mol H} \cdot \frac{2.0158 \text{ g}}{\text{mol}} = 419{,}900 \text{ g} \cdot \frac{1 \text{ kg}}{1000 \text{ g}} = 419.9 \text{ kg}$$

Rounding this value to 3 sig digs requires expressing it in scientific notation, giving 4.20×10^2 kg.

15. a.

$$V = 2.000 \text{ L}$$

$$T = 27.0°C \rightarrow 27.0°C + 273.2 = 300.2 \text{ K}$$

$$P = 23.11 \text{ atm}$$

$$R = 0.08206 \frac{\text{L·atm}}{\text{mol·K}}$$

$$PV = nRT \rightarrow n = \frac{PV}{RT} = \frac{23.11 \text{ atm} \cdot 2.000 \text{ L}}{0.08206 \dfrac{\text{L·atm}}{\text{mol·K}} \cdot 300.2 \text{ K}} = 1.876 \text{ mol}$$

15. b.

$$\frac{133.0 \text{ g}}{1.876 \text{ mol}} = 70.90 \frac{\text{g}}{\text{mol}}$$

All halogens are diatomic, which means the above molecular mass is twice the atomic mass. Dividing by 2 we get 35.45 g/mol, which is the atomic mass of chlorine.

16. a.

$V = 275 \text{ L}$

$T = 29°\text{C} \rightarrow 29°\text{C} + 273.2 = 302.2 \text{ K}$

$P = 765 \text{ mm Hg} = 765 \text{ Torr}$

$R = 62.36 \; \dfrac{\text{L} \cdot \text{Torr}}{\text{mol} \cdot \text{K}}$

$PV = nRT \rightarrow n = \dfrac{PV}{RT} = \dfrac{765 \text{ Torr} \cdot 275 \text{ L}}{62.36 \; \dfrac{\text{L} \cdot \text{Torr}}{\text{mol} \cdot \text{K}} \cdot 302.2 \text{ K}} = 11.16 \text{ mol}$

Rounding this value to 2 sig digs gives 11.2 mol.

16. b.

At the new altitude, the number of moles of helium will be the same. We will use the value with the extra sig dig and round off to 3 sig digs at the end.

$n = 11.16 \text{ mol}$

$T = 221 \text{ K}$

$P = 25.0 \text{ kPa}$

$R = 8.314 \; \dfrac{\text{L} \cdot \text{kPa}}{\text{mol} \cdot \text{K}}$

$PV = nRT \rightarrow V = \dfrac{nRT}{P} = \dfrac{11.16 \text{ mol} \cdot 8.314 \; \dfrac{\text{L} \cdot \text{kPa}}{\text{mol} \cdot \text{K}} \cdot 221 \text{ K}}{25.0 \text{ kPa}} = 820 \text{ L}$

In order to show 3 sig digs, we write this value as 8.20×10^2 L.

17.

In this problem, only T and P and variables. Write the ideal gas law with these variables on one side and all the constants on the other side:

$PV = nRT \rightarrow \dfrac{P}{T} = \dfrac{nR}{V} = k$

Since P/T is a constant, P/T at one set of conditions is equal to P/T at any other set of conditions. So we can write

$\dfrac{P_1}{T_1} = \dfrac{P_2}{T_2}$

We must work only with absolute measurement values for P and T.

$P_{abs} = P_{gauge} + P_{atm}$

$P_1 = 35 \text{ psig} + 14.7 \text{ psig} = 49.7 \text{ psia}$

$T_1 = 36°\text{C} \rightarrow 36°\text{C} + 273.2 = 309.2 \text{ K}$

$T_2 = -5°\text{C} \rightarrow -5°\text{C} + 273.2 = 268.2 \text{ K}$

$\dfrac{P_1}{T_1} = \dfrac{P_2}{T_2} \rightarrow P_2 = P_1 \cdot \dfrac{T_2}{T_1} = 49.7 \text{ psia} \cdot \dfrac{268.2 \text{ K}}{309.2 \text{ K}} = 43.1 \text{ psia}$

Finally, we must convert this result back to a gauge pressure with 2 sig digs.

$P_{abs} = P_{gauge} + P_{atm}$

$P_{gauge} = P_{abs} - P_{atm} = 43.1 \text{ psia} - 14.7 \text{ psi} = 28 \text{ psig}$

---

18. a.

As in the previous problem, the pressure must be converted to an absolute pressure.

$V = 12.00 \text{ L}$

$P = 119.0 \text{ bar (g)} \rightarrow 119.0 \text{ bar} + 1.01325 \text{ bar} = 120.01 \text{ bar (abs)} \cdot \dfrac{100 \text{ kPa}}{1 \text{ bar}} = 12{,}001 \text{ kPa}$

$T = 22.00°C \rightarrow 22.00°C + 273.15 = 295.15 \text{ K}$

$R = 8.314 \dfrac{\text{L} \cdot \text{kPa}}{\text{mol} \cdot \text{K}}$

$PV = nRT \rightarrow n = \dfrac{PV}{RT} = \dfrac{12{,}001 \text{ kPa} \cdot 12.00 \text{ L}}{8.314 \dfrac{\text{L} \cdot \text{kPa}}{\text{mol} \cdot \text{K}} \cdot 295.15 \text{ K}} = 58.688 \text{ mol}$

Rounding to 4 sig digs we have 58.69 mol.

---

18. b.

$n = 58.688 \text{ mol}$

$P = 100 \text{ kPa}$

$T = 0.0°C \rightarrow 0.0°C + 273.15 = 273.15 \text{ K}$

$R = 8.314 \dfrac{\text{L} \cdot \text{kPa}}{\text{mol} \cdot \text{K}}$

$PV = nRT \rightarrow V = \dfrac{nRT}{P} = \dfrac{58.688 \text{ mol} \cdot 8.314 \dfrac{\text{L} \cdot \text{kPa}}{\text{mol} \cdot \text{K}} \cdot 273.15 \text{ K}}{100 \text{ kPa}} = 1333 \text{ L}$

---

19. a.

The molar mass is the mass divided by the number of moles (g/mol).

$V = 0.975 \text{ L}$

$T = 26°C \rightarrow 26°C + 273.2 = 299.2 \text{ K}$

$P = 868 \text{ Torr}$

$R = 62.36 \dfrac{\text{L} \cdot \text{Torr}}{\text{mol} \cdot \text{K}}$

$PV = nRT \rightarrow n = \dfrac{PV}{RT} = \dfrac{868 \text{ Torr} \cdot 0.975 \text{ L}}{62.36 \dfrac{\text{L} \cdot \text{Torr}}{\text{mol} \cdot \text{K}} \cdot 299.2 \text{ K}} = 0.04535 \text{ mol}$

molar mass: $\dfrac{0.942 \text{ g}}{0.04535 \text{ mol}} = 20.8 \dfrac{\text{g}}{\text{mol}}$

---

19. b.

$$V = 888 \text{ mL} \cdot \frac{1 \text{ L}}{1000 \text{ mL}} = 0.888 \text{ L}$$

$$T = -32°C \rightarrow -32°C + 273.2 = 241.2 \text{ K}$$

$$P = 2.14 \text{ bar} \cdot \frac{100 \text{ kPa}}{1 \text{ bar}} = 214 \text{ kPa}$$

$$R = 8.314 \frac{\text{L} \cdot \text{kPa}}{\text{mol} \cdot \text{K}}$$

$$PV = nRT \rightarrow n = \frac{PV}{RT} = \frac{214 \text{ kPa} \cdot 0.888 \text{ L}}{8.314 \dfrac{\text{L} \cdot \text{kPa}}{\text{mol} \cdot \text{K}} \cdot 241.2 \text{ K}} = 0.09476 \text{ mol}$$

$$\text{molar mass: } \frac{0.651 \text{ g}}{0.09476 \text{ mol}} = 6.87 \frac{\text{g}}{\text{mol}}$$

23.

$$X_{N_2} = 0.820$$

$$X_{Ar} = 0.120$$

$$X_{CH_4} = 0.060$$

$$P_T = 1.61 \text{ atm}$$

$$P_{N_2} = X_{N_2} \cdot P_T = 0.820 \cdot 1.61 \text{ atm} = 1.32 \text{ atm}$$

$$P_{Ar} = X_{Ar} \cdot P_T = 0.120 \cdot 1.61 \text{ atm} = 0.193 \text{ atm}$$

$$P_{CH_4} = X_{CH_4} \cdot P_T = 0.060 \cdot 1.61 \text{ atm} = 0.097 \text{ atm}$$

---

24. a.

$$X_{Ar} = 0.805$$

$$X_{O_2} = 0.180$$

$$X_{CO_2} = 0.015$$

$$P_T = 748 \text{ Torr}$$

$$P_{Ar} = X_{Ar} \cdot P_T = 0.805 \cdot 748 \text{ Torr} = 602.1 \text{ Torr}$$

For 3 sig digs, this rounds to 602 Torr.

---

24. b.

$$P = 602.1 \text{ Torr}$$

$$V = 165 \text{ L}$$

$$T = 295 \text{ K}$$

$$R = 62.36 \frac{\text{L} \cdot \text{Torr}}{\text{mol} \cdot \text{K}}$$

$$PV = nRT \rightarrow n = \frac{PV}{RT} = \frac{602.1 \text{ Torr} \cdot 165 \text{ L}}{62.36 \dfrac{\text{L} \cdot \text{Torr}}{\text{mol} \cdot \text{K}} \cdot 295 \text{ K}} = 5.40 \text{ mol}$$

25.

The partial pressure of propane is given. From this and the number of moles of each gas the total pressure may be determined:

$$P_{C_3H_8} = 2.55 \text{ bar}$$

$$m_{C_3H_8} = 2.55 \text{ g}$$

$$m_{CH_4} = 1.01 \text{ g}$$

$$n_{C_3H_8} = 2.55 \text{ g } C_3H_8 \cdot \frac{\text{mol}}{44.096 \text{ g}} = 0.05783 \text{ mol}$$

$$n_{CH_4} = 1.01 \text{ g } CH_4 \cdot \frac{\text{mol}}{16.043 \text{ g}} = 0.06296 \text{ mol}$$

$$n_T = n_{C_3H_8} + n_{CH_4} = 0.05783 \text{ mol} + 0.06296 \text{ mol} = 0.1208 \text{ mol}$$

$$P_{C_3H_8} = \frac{n_{C_3H_8}}{n_T} \cdot P_T \rightarrow P_T = P_{C_3H_8} \cdot \frac{n_T}{n_{C_3H_8}} = 2.55 \text{ bar} \cdot \frac{0.1208 \text{ mol}}{0.05783 \text{ mol}} = 5.327 \text{ bar}$$

Rounding to 3 sig digs we have $P_T = 5.33$ bar. With the total pressure we can determine the partial pressure of methane:

$$P_{CH_4} = \frac{n_{CH_4}}{n_T} \cdot P_T = \frac{0.06296 \text{ mol}}{0.1208 \text{ mol}} \cdot 5.327 \text{ bar} = 2.78 \text{ bar}$$

Finally, the mole fractions are:

$$X_{C_3H_8} = \frac{n_{C_3H_8}}{n_T} = \frac{0.05783 \text{ mol}}{0.1208 \text{ mol}} = 0.479$$

$$X_{CH_4} = \frac{n_{CH_4}}{n_T} = \frac{0.06296 \text{ mol}}{0.1208 \text{ mol}} = 0.521$$

26. a.

O_2:

$V = 2.00$ L

$P = 1.55$ atm

$T = 23°C \rightarrow 23°C + 273.2 = 296.2$ K

$R = 0.08206 \dfrac{\text{L·atm}}{\text{mol·K}}$

$PV = nRT \rightarrow n = \dfrac{PV}{RT} = \dfrac{1.55 \text{ atm·} 2.00 \text{ L}}{0.08206 \dfrac{\text{L·atm}}{\text{mol·K}} \cdot 296.2 \text{ K}} = 0.1275$ mol

Rounding this value to 3 sig digs gives 0.128 mol.

H_2:

$V = 1.00$ L

$P = 1.05$ atm

$T = 23°C \rightarrow 23°C + 273.2 = 296.2$ K

$R = 0.08206 \dfrac{\text{L·atm}}{\text{mol·K}}$

$PV = nRT \rightarrow n = \dfrac{PV}{RT} = \dfrac{1.05 \text{ atm·} 1.00 \text{ L}}{0.08206 \dfrac{\text{L·atm}}{\text{mol·K}} \cdot 296.2 \text{ K}} = 0.04320$ mol

Rounding this value to 3 sig digs gives 0.0432 mol.

26. b.

$V = 3.00$ L

$n_{O_2} = 0.1275$ mol

$T = 23°C \rightarrow 23°C + 273.2 = 296.2$ K

$R = 0.08206 \dfrac{\text{L·atm}}{\text{mol·K}}$

$PV = nRT \rightarrow P_{O_2} = \dfrac{n_{O_2} RT}{V} = \dfrac{0.1275 \text{ mol·} 0.08206 \dfrac{\text{L·atm}}{\text{mol·K}} \cdot 296.2 \text{ K}}{3.00 \text{ L}} = 1.033$ atm

Rounding this value to 3 sig digs gives 1.03 atm.

$V = 3.00$ L

$n_{H_2} = 0.04320$ mol

$T = 23°C \rightarrow 23°C + 273.2 = 296.2$ K

$R = 0.08206 \dfrac{\text{L·atm}}{\text{mol·K}}$

$PV = nRT \rightarrow P_{H_2} = \dfrac{n_{H_2} RT}{V} = \dfrac{0.04320 \text{ mol·} 0.08206 \dfrac{\text{L·atm}}{\text{mol·K}} \cdot 296.2 \text{ K}}{3.00 \text{ L}} = 0.3500$ atm

Rounding this value to 3 sig digs gives 0.350 atm.

26. c.

$$P_T = P_{O_2} + P_{H_2} = 1.033 \text{ atm} + 0.3500 \text{ atm} = 1.38 \text{ atm}$$

---

27. a/b.

First, determine the partial pressure of water at this temperature, which is its vapor pressure. From Table A.4, this is 2.6453 kPa.

$$P_{water} = 2.6453 \text{ kPa} \cdot \frac{1 \text{ atm}}{101.325 \text{ kPa}} = 0.02611 \text{ atm}$$

$$P_T = 1.00 \text{ atm}$$

$$P_T = P_{water} + P_{N_2} \rightarrow P_{N_2} = P_T - P_{water} = 1.00 \text{ atm} - 0.02611 \text{ atm} = 0.974 \text{ atm}$$

The addition rule requires us to round this partial pressure to the second decimal place, giving 0.97 atm. Now we calculate the number of moles using the ideal gas equation:

$$P_{N_2} = 0.974 \text{ atm}$$

$$V = 285 \text{ mL} \cdot \frac{1 \text{ L}}{1000 \text{ mL}} = 0.285 \text{ L}$$

$$T = 22°C \rightarrow 22°C + 273.2 \text{ K} = 295.2 \text{ K}$$

$$R = 0.08206 \frac{\text{L} \cdot \text{atm}}{\text{mol} \cdot \text{K}}$$

$$PV = nRT \rightarrow n_{N_2} = \frac{P_{N_2} V}{RT} = \frac{0.974 \text{ atm} \cdot 0.285 \text{ L}}{0.08206 \frac{\text{L} \cdot \text{atm}}{\text{mol} \cdot \text{K}} \cdot 295.2 \text{ K}} = 0.01146 \text{ mol}$$

With 2 sig digs, this is 0.011 mol. To obtain the mole fraction, we have to first obtain the number of moles of water vapor:

$$P_{water} = 0.02611 \text{ atm}$$

$$V = 0.285 \text{ L}$$

$$T = 295.2 \text{ K}$$

$$R = 0.08206 \frac{\text{L} \cdot \text{atm}}{\text{mol} \cdot \text{K}}$$

$$PV = nRT \rightarrow n_{water} = \frac{P_{water} V}{RT} = \frac{0.02611 \text{ atm} \cdot 0.285 \text{ L}}{0.08206 \frac{\text{L} \cdot \text{atm}}{\text{mol} \cdot \text{K}} \cdot 295.2 \text{ K}} = 0.0003072 \text{ mol}$$

This value still has 3 sig digs, so it rounds to 0.000307 mol. Now we can calculate the total number of moles and the mole fractions.

$$n_T = n_{N_2} + n_{water} = 0.01146 \text{ mol} + 0.0003072 \text{ mol} = 0.01177 \text{ mol}$$

$$X_{N_2} = \frac{n_{N_2}}{n_T} = \frac{0.01146 \text{ mol}}{0.01177 \text{ mol}} = 0.974$$

$$X_{water} = \frac{n_{water}}{n_T} = \frac{0.0003072 \text{ mol}}{0.01177 \text{ mol}} = 0.0261$$

The value for the moles of N_2 limits precision to the third decimal place in n_T, which in turn limits the precision of these mole fractions to 2 sig digs, giving 0.97 and 0.026.

28.

We need to use the partial pressure of H_2 to determine the number of moles of H_2. From this we can do the stoichiometry to determine the amount of zinc consumed. From Table A.4, the vapor pressure of water at these conditions is 28.38 Torr.

$$P_T = P_{H_2} + P_{water} \rightarrow P_{H_2} = P_T - P_{water} = 757 \text{ Torr} - 28.38 \text{ Torr} = 728.6 \text{ Torr}$$

$$V = 184 \text{ mL} \cdot \frac{1 \text{ L}}{1000 \text{ mL}} = 0.184 \text{ L}$$

$$T = 28°C \rightarrow 28°C + 273.2 = 301.2 \text{ K}$$

$$R = 62.36 \frac{\text{L} \cdot \text{Torr}}{\text{mol} \cdot \text{K}}$$

$$PV = nRT \rightarrow n_{H_2} = \frac{P_{H_2}V}{RT} = \frac{728.6 \text{ Torr} \cdot 0.184 \text{ L}}{62.36 \frac{\text{L} \cdot \text{Torr}}{\text{mol} \cdot \text{K}} \cdot 301.2 \text{ K}} = 0.007138 \text{ mol}$$

$$0.007138 \text{ mol } H_2 \cdot \frac{1 \text{ mol Zn}}{1 \text{ mol } H_2} = 0.007138 \text{ mol Zn}$$

$$0.007138 \text{ mol Zn} \cdot \frac{65.39 \text{ g}}{\text{mol}} = 0.467 \text{ g}$$

29.

$$2HgO \rightarrow 2Hg + O_2$$

From Table A.4, the vapor pressure of water at these conditions is 15.49 Torr.

$$P_T = P_{O_2} + P_{water} \rightarrow P_{O_2} = P_T - P_{water} = 725 \text{ Torr} - 15.49 \text{ Torr} = 709.5 \text{ Torr}$$

$$V = 355 \text{ mL} \cdot \frac{1 \text{ L}}{1000 \text{ mL}} = 0.355 \text{ L}$$

$$T = 18°C \rightarrow 18°C + 273.2 = 291.2 \text{ K}$$

$$R = 62.36 \frac{\text{L} \cdot \text{Torr}}{\text{mol} \cdot \text{K}}$$

$$PV = nRT \rightarrow n_{O_2} = \frac{P_{O_2}V}{RT} = \frac{709.5 \text{ Torr} \cdot 0.355 \text{ L}}{62.36 \frac{\text{L} \cdot \text{Torr}}{\text{mol} \cdot \text{K}} \cdot 291.2 \text{ K}} = 0.01387 \text{ mol}$$

$$0.01387 \text{ mol } O_2 \cdot \frac{2 \text{ mol HgO}}{1 \text{ mol } O_2} = 0.02774 \text{ mol HgO}$$

$$0.02774 \text{ mol HgO} \cdot \frac{216.6 \text{ g}}{\text{mol}} = 6.01 \text{ g}$$

30.

$$2CO + O_2 \rightarrow 2CO_2$$

$$5.0 \text{ L CO} \cdot \frac{1 \text{ mol } O_2}{2 \text{ mol CO}} = 2.5 \text{ L } O_2$$

$$5.0 \text{ L CO} \cdot \frac{2 \text{ mol } CO_2}{2 \text{ mol CO}} = 5.0 \text{ L } CO_2$$

31.

$$C_3H_8 + 5O_2 \rightarrow 3CO_2 + 4H_2O$$

$$30.00 \text{ m}^3 \text{ C}_3\text{H}_8 \cdot \frac{5 \text{ mol O}_2}{1 \text{ mol C}_3\text{H}_8} = 150.0 \text{ m}^3 \text{ O}_2$$

$$30.00 \text{ m}^3 \text{ C}_3\text{H}_8 \cdot \frac{3 \text{ mol CO}_2}{1 \text{ mol C}_3\text{H}_8} = 90.00 \text{ m}^3 \text{ CO}_2$$

$$30.00 \text{ m}^3 \text{ C}_3\text{H}_8 \cdot \frac{4 \text{ mol H}_2\text{O}}{1 \text{ mol C}_3\text{H}_8} = 120.0 \text{ m}^3 \text{ H}_2\text{O}$$

32.

$$2Fe(OH)_3 \rightarrow Fe_2O_3 + 3H_2O$$

$V = 2.00 \text{ L}$

$P = 101.325 \text{ kPa}$

$T = 390 \text{ K}$

$$R = 8.314 \frac{L \cdot kPa}{mol \cdot K}$$

$$PV = nRT \rightarrow n = \frac{PV}{RT} = \frac{101.325 \text{ kPa} \cdot 2.00 \text{ L}}{8.314 \dfrac{L \cdot kPa}{mol \cdot K} \cdot 390 \text{ K}} = 0.0625 \text{ mol H}_2\text{O}$$

$$0.0625 \text{ mol H}_2\text{O} \cdot \frac{2 \text{ mol Fe(OH)}_3}{3 \text{ mol H}_2\text{O}} = 0.0417 \text{ mol Fe(OH)}_3$$

$$0.0417 \text{ mol Fe(OH)}_3 \cdot \frac{106.9 \text{ g}}{mol} = 4.5 \text{ g Fe(OH)}_3$$

$$0.0625 \text{ mol H}_2\text{O} \cdot \frac{1 \text{ mol Fe}_2\text{O}_3}{3 \text{ mol H}_2\text{O}} = 0.0208 \text{ mol Fe}_2\text{O}_3$$

$$0.0208 \text{ mol Fe}_2\text{O}_3 \cdot \frac{159.7 \text{ g}}{mol} = 3.3 \text{ g Fe}_2\text{O}_3$$

33.

$V = 31,150 \text{ L}$

$P = 101.325 \text{ kPa}$

$T = 24°C \rightarrow 24°C + 273.2 = 297.2 \text{ K}$

$$R = 8.314 \frac{L \cdot kPa}{mol \cdot K}$$

$$PV = nRT \rightarrow n = \frac{PV}{RT} = \frac{101.325 \text{ kPa} \cdot 31,150 \text{ L}}{8.314 \dfrac{L \cdot kPa}{mol \cdot K} \cdot 297.2 \text{ K}} = 1277 \text{ mol H}_2$$

$$1277 \text{ mol H}_2 \cdot \frac{1 \text{ mol Fe}}{1 \text{ mol H}_2} = 1277 \text{ mol Fe}$$

$$1277 \text{ mol Fe} \cdot \frac{55.847 \text{ g}}{mol} = 71,400 \text{ g Fe} \cdot \frac{1 \text{ kg}}{1000 \text{ g}} = 71.4 \text{ kg}$$

Note that although the temperature was given with 2 sig digs, this value has 3 sig digs after being converted to kelvins. Thus, the result has 3 sig digs.

34.

$$Fe + H_2SO_4 \rightarrow FeSO_4 + H_2$$

$$1277 \text{ mol } H_2 \cdot \frac{1 \text{ mol Fe}}{1 \text{ mol } H_2} = 1277 \text{ mol Fe}$$

$$1277 \text{ mol Fe} \cdot \frac{55.847 \text{ g}}{\text{mol}} = 71,400 \text{ g Fe} \cdot \frac{1 \text{ kg}}{1000 \text{ g}} = 71.4 \text{ kg}$$

35.

$$Mg + 2HCl \rightarrow MgCl_2 + H_2$$

$$V = 0.015 \text{ L}$$

$$P = 101.325 \text{ kPa}$$

$$T = 273.15 \text{ K}$$

$$R = 8.314 \frac{\text{L} \cdot \text{kPa}}{\text{mol} \cdot \text{K}}$$

$$PV = nRT \rightarrow n = \frac{PV}{RT} = \frac{101.325 \text{ kPa} \cdot 0.015 \text{ L}}{8.314 \dfrac{\text{L} \cdot \text{kPa}}{\text{mol} \cdot \text{K}} \cdot 273.15 \text{ K}} = 6.69 \times 10^{-4} \text{ mol } H_2$$

$$6.69 \times 10^{-4} \text{ mol } H_2 \cdot \frac{1 \text{ mol Mg}}{1 \text{ mol } H_2} = 6.69 \times 10^{-4} \text{ mol Mg}$$

$$6.69 \times 10^{-4} \text{ mol Mg} \cdot \frac{24.305 \text{ g}}{\text{mol}} = 0.016 \text{ g}$$

36.

From Table A.4, the vapor pressure of water at these conditions is 17.55 Torr.

$$0.883 \text{ g CaC}_2 \cdot \frac{\text{mol}}{64.10 \text{ g}} = 0.01378 \text{ mol CaC}_2$$

$$P_T = P_{C_2H_2} + P_{water} \rightarrow P_{C_2H_2} = P_T - P_{water} = 735 \text{ Torr} - 17.55 \text{ Torr} = 717.5 \text{ Torr}$$

$$T = 20°C \rightarrow 20°C + 273.2 = 293.2 \text{ K}$$

$$R = 62.36 \frac{\text{L} \cdot \text{Torr}}{\text{mol} \cdot \text{K}}$$

$$PV = nRT \rightarrow V_{C_2H_2} = \frac{n_{C_2H_2} RT}{P_{C_2H_2}} = \frac{0.01378 \text{ mol} \cdot 62.36 \dfrac{\text{L} \cdot \text{Torr}}{\text{mol} \cdot \text{K}} \cdot 293.2 \text{ K}}{717.5 \text{ Torr}} = 0.351 \text{ L}$$

37.

$2KI + Cl_2 \rightarrow 2KCl + I_2$

37. a.

$V = 7.75$ L

$P = 100$ kPa

$T = 273.15$ K

$R = 8.314 \dfrac{\text{L} \cdot \text{kPa}}{\text{mol} \cdot \text{K}}$

$PV = nRT \rightarrow n = \dfrac{PV}{RT} = \dfrac{100 \text{ kPa} \cdot 7.75 \text{ L}}{8.314 \dfrac{\text{L} \cdot \text{kPa}}{\text{mol} \cdot \text{K}} \cdot 273.15 \text{ K}} = 0.3413 \text{ mol I}_2$

$0.3413 \text{ mol I}_2 \cdot \dfrac{2 \text{ mol KI}}{1 \text{ mol I}_2} = 0.6826 \text{ mol KI}$

$0.3413 \text{ mol I}_2 \cdot \dfrac{1 \text{ mol Cl}_2}{1 \text{ mol I}_2} = 0.3413 \text{ mol Cl}_2$

$0.3413 \text{ mol I}_2 \cdot \dfrac{2 \text{ mol KCl}}{1 \text{ mol I}_2} = 0.6826 \text{ mol KCl}$

Rounding these to 3 sig digs gives 0.341 mol I_2, 0.683 mol KI, 0.341 mol Cl_2, and 0.683 mol KCl.

37. b.

$0.3413 \text{ mol I}_2 \cdot \dfrac{253.8 \text{ g}}{\text{mol}} = 86.6 \text{ g I}_2$

$0.6826 \text{ mol KI} \cdot \dfrac{165.2 \text{ g}}{\text{mol}} = 113 \text{ g KI}$

$0.3413 \text{ mol Cl}_2 \cdot \dfrac{70.91 \text{ g}}{\text{mol}} = 24.2 \text{ g Cl}_2$

$0.6286 \text{ mol KCl} \cdot \dfrac{74.55 \text{ g}}{\text{mol}} = 50.9 \text{ g KCl}$

38.

$2Al + 6HCl \rightarrow 2AlCl_3 + 3H_2$

38. a.

$110.0 \text{ g HCl} \cdot \dfrac{\text{mol}}{36.4606 \text{ g}} = 3.017 \text{ mol HCl}$

$25.00 \text{ g Al} \cdot \dfrac{\text{mol}}{26.9815 \text{ g}} = 0.92656 \text{ mol Al}$

$0.92656 \text{ mol Al} \cdot \dfrac{6 \text{ mol HCl}}{2 \text{ mol Al}} = 2.78 \text{ mol HCl}$

More HCl than this is available, so Al is the limiting reactant.

38. b.

$$0.92656 \text{ mol Al} \cdot \frac{3 \text{ mol H}_2}{2 \text{ mol Al}} = 1.3898 \text{ mol H}_2$$

$P = 101.325 \text{ kPa}$

$T = 32.0°C \rightarrow 32.0°C + 273.15 = 305.15 \text{ K}$

$$R = 8.314 \frac{\text{L} \cdot \text{kPa}}{\text{mol} \cdot \text{K}}$$

$$PV = nRT \rightarrow V = \frac{nRT}{P} = \frac{1.3898 \text{ mol} \cdot 8.314 \dfrac{\text{L} \cdot \text{kPa}}{\text{mol} \cdot \text{K}} \cdot 305.15 \text{ K}}{101.325 \text{ kPa}} = 34.80 \text{ L}$$

39.

$$CO + 2H_2 \rightarrow CH_3OH$$

39. a.

$$660.0 \text{ m}^3 \text{ CO} \cdot \frac{2 \text{ m}^3 \text{ H}_2}{1 \text{ m}^3 \text{ CO}} = 1320 \text{ m}^3 \text{ H}_2$$

This much H_2 is not available. Thus, H_2 is the limiting reactant and CO is present in excess.

---

39. b.

$$1210.0 \text{ m}^3 \text{ H}_2 \cdot \frac{1 \text{ m}^3 \text{ CO}}{2 \text{ m}^3 \text{ H}_2} = 605.0 \text{ m}^3 \text{ CO}$$

$660.0 \text{ m}^3 - 605.0 \text{ m}^3 = 55.0 \text{ m}^3$

---

39. c.

$$1210.0 \text{ m}^3 \text{ H}_2 \cdot \frac{1 \text{ m}^3 \text{ CH}_3OH}{2 \text{ m}^3 \text{ H}_2} = 605.0 \text{ m}^3 \text{ CH}_3OH$$

40.

$$4C_3H_5N_3O_9 \rightarrow 10H_2O + 12CO_2 + 6N_2 + O_2$$

40. a.

$$1.00 \text{ g H}_2\text{O} \cdot \frac{\text{mol}}{18.015 \text{ g}} = 0.05551 \text{ mol H}_2\text{O}$$

$$0.05551 \text{ mol H}_2\text{O} \cdot \frac{12 \text{ mol CO}_2}{10 \text{ mol H}_2\text{O}} = 0.06661 \text{ mol CO}_2$$

$$0.05551 \text{ mol H}_2\text{O} \cdot \frac{6 \text{ mol N}_2}{10 \text{ mol H}_2\text{O}} = 0.03331 \text{ mol N}_2$$

$$0.05551 \text{ mol H}_2\text{O} \cdot \frac{1 \text{ mol O}_2}{10 \text{ mol H}_2\text{O}} = 0.005551 \text{ mol O}_2$$

$$n_T = 0.05551 \text{ mol} + 0.06661 \text{ mol} + 0.03331 \text{ mol} + 0.00555 \text{ mol} = 0.16098 \text{ mol}$$

$$P = 50.0 \text{ bar} \cdot \frac{100 \text{ kPa}}{1 \text{ bar}} = 5000.0 \text{ kPa}$$

$$T = 350°C \rightarrow 350°C + 273 = 623 \text{ K}$$

$$R = 8.314 \frac{\text{L} \cdot \text{kPa}}{\text{mol} \cdot \text{K}}$$

$$PV = nRT \rightarrow V = \frac{nRT}{P} = \frac{0.16098 \text{ mol} \cdot 8.314 \frac{\text{L} \cdot \text{kPa}}{\text{mol} \cdot \text{K}} \cdot 623 \text{ K}}{5000.0 \text{ kPa}} = 0.17 \text{ L}$$

40. b.

$$0.05551 \text{ mol H}_2\text{O} \cdot \frac{4 \text{ mol C}_3\text{H}_5\text{N}_3\text{O}_9}{10 \text{ mol H}_2\text{O}} = 0.02220 \text{ mol C}_3\text{H}_5\text{N}_3\text{O}_9$$

$$0.02220 \text{ mol C}_3\text{H}_5\text{N}_3\text{O}_9 \cdot \frac{227.1 \text{ g}}{\text{mol}} = 5.0 \text{ g}$$

41.

$$E = 3.20 \times 10^{-18} \text{ J}$$

$$E = \frac{hv}{\lambda} \rightarrow \lambda = \frac{hv}{E} = \frac{6.626 \times 10^{-34} \text{ J} \cdot \text{s} \cdot 2.9979 \times 10^8 \frac{\text{m}}{\text{s}}}{3.20 \times 10^{-18} \text{ J}} = 6.21 \times 10^{-8} \text{ m} \cdot \frac{1 \times 10^9 \text{ nm}}{1 \text{ m}} = 62.1 \text{ nm}$$

This wavelength is in the UV region.

42.

$$\frac{3 \cdot 12.011}{227.087} = 0.15867 \rightarrow 15.867\% \text{ C}$$

$$\frac{5 \cdot 1.0079}{227.087} = 0.022192 \rightarrow 2.2192\% \text{ H}$$

$$\frac{3 \cdot 14.0067}{227.087} = 0.185040 \rightarrow 18.5040\% \text{ N}$$

$$\frac{9 \cdot 15.9994}{227.087} = 0.634094 \rightarrow 63.4094\% \text{ O}$$

43. a.

$$53.5 \text{ kg} \cdot \frac{1000 \text{ g}}{1 \text{ kg}} = 53{,}500 \text{ g Fe}$$

$$53{,}500 \text{ g Fe} \cdot \frac{\text{mol}}{55.847 \text{ g}} = 958.0 \text{ mol Fe}$$

$$958.0 \text{ mol Fe} \cdot \frac{1 \text{ mol H}_2}{1 \text{ mol Fe}} = 958.0 \text{ mol H}_2$$

$$V = 19{,}625 \text{ L}$$

$$P = 101.325 \text{ kPa}$$

$$T = 24°\text{C} \rightarrow 24°\text{C} + 273.2 = 297.2 \text{ K}$$

$$R = 8.314 \frac{\text{L} \cdot \text{kPa}}{\text{mol} \cdot \text{K}}$$

$$PV = nRT \rightarrow n = \frac{PV}{RT} = \frac{101.325 \text{ kPa} \cdot 19{,}625 \text{ L}}{8.314 \dfrac{\text{L} \cdot \text{kPa}}{\text{mol} \cdot \text{K}} \cdot 297.2 \text{ K}} = 804.7 \text{ mol}$$

$$\frac{804.7 \text{ mol}}{958.0 \text{ mol}} = 84.0\%$$

43. b.

$$804.7 \text{ mol} \cdot \frac{6.022 \times 10^{23} \text{ particles}}{\text{mol}} = 4.85 \times 10^{26} \text{ particles (molecules)}$$

43. c.

$$958.0 \text{ mol Fe} \cdot \frac{1 \text{ mol FeCl}_2}{1 \text{ mol Fe}} = 958.0 \text{ mol FeCl}_2$$

$$958.0 \text{ mol FeCl}_2 \cdot \frac{126.8 \text{ g}}{\text{mol}} = 121{,}000 \text{ g} \cdot \frac{1 \text{ kg}}{1000 \text{ g}} = 121 \text{ kg}$$

45.

$$C_{ice} = 0.0364 \ \frac{kJ}{mol \cdot K}$$

$$C_{water} = 0.0752 \ \frac{kJ}{mol \cdot K}$$

$$H_f = 6.01 \ \frac{kJ}{mol}$$

$$m = 22.0 \ kg \cdot \frac{1000 \ g}{1 \ kg} = 22,000 \ g$$

$$n = 22,000 \ g \cdot \frac{mol}{18.015 \ g} = 1221 \ mol$$

$$\Delta T = T_f - T_i = 20.0°C - 0.0°C = 20.0°C = 20.0 \ K$$

$$Q_{melt} = nH_f = 1221 \ mol \cdot 6.01 \ \frac{kJ}{mol} = 7340 \ kJ$$

$$Q_{warm} = Cn\Delta T = 0.0752 \ \frac{kJ}{mol \cdot K} \cdot 1221 \ mol \cdot 20.0 \ K = 1840 \ kJ$$

$$Q_{total} = Q_{melt} + Q_{warm} = 7340 \ kJ + 1840 \ kJ = 9180 \ kJ$$

Chapter 10

15. d.

$$\frac{0.7 - 0.54}{0.7} \cdot 100\% = 23\%$$

16. b

$$\frac{30.7 + 32.6}{2} = 31.7$$

16. c

$m_{solution} = 250.0$ g

solubility (mass %) = 34.9

$0.349 \cdot 250.0$ g $= 87.3$ g

16.d

$m_{solution} = 45$ g

$Al_2(SO_4)_3$ solubility (mass percent) = 28.2

$0.282 \cdot 45$ g $= 12.69$ g $Al_2(SO_4)_3$

12.69 g $\cdot \dfrac{1 \text{ mol}}{342.1538 \text{ g}} = 0.0371$ mol $Al_2(SO_4)_3$

H_2O (mass percent) = 71.8

$0.718 \cdot 45$ g $= 32.31$ g H_2O

32.31 g $\cdot \dfrac{1 \text{ mol}}{18.015 \text{ g}} = 1.793$ mol H_2O

total moles $= 0.0371$ mol $Al_2(SO_4)_3 + 1.793$ mol $H_2O = 1.8301$ mol solution

mol % $Al_2(SO_4)_3$:

$\dfrac{0.0371}{1.8301} \times 100\% = 2.0\%$

mol % H_2O:

$\dfrac{1.793}{1.8301} \times 100\% = 98\%$

17. a.

$8.00 \ \dfrac{\text{mol}}{\text{L}} \cdot 2.25 \text{ L} = 18.0$ mol

18.0 mol NaOH $\cdot \dfrac{40.00 \text{ g}}{\text{mol}} = 719.9$ g

To round this to 3 sig digs, write it in scientific notation: 7.20×10^2 g.

17. b.

$$15.0 \ \frac{\text{mol}}{\text{L}} \cdot 1.00 \ \text{L} = 15.0 \ \text{mol}$$

$$15.0 \ \text{mol} \ CH_3COOH \cdot \frac{60.05 \ \text{g}}{\text{mol}} = 901 \ \text{g}$$

18.

$$65.11 \ \text{g} \ (NH_4)_2 SO_4 \cdot \frac{\text{mol}}{132.140 \ \text{g}} = 0.4927 \ \text{mol}$$

$$125 \ \text{mL} \cdot \frac{1 \ \text{L}}{1000 \ \text{mL}} = 0.125 \ \text{L}$$

$$\frac{0.4927 \ \text{mol}}{0.125 \ \text{L}} = 3.94 \ M$$

19.

$$1.25 \ \frac{\text{mol}}{\text{L}} \cdot \frac{164.1 \ \text{g}}{\text{mol}} = 205.1 \ \frac{\text{g}}{\text{L}}$$

$$205.1 \ \frac{\text{g}}{\text{L}} \cdot 3.10 \ \text{L} = 636 \ \text{g}$$

20.

$$1.2 \ \frac{\text{mol}}{\text{L}} \cdot \frac{169.87 \ \text{g}}{\text{mol}} = 203.8 \ \frac{\text{g}}{\text{L}}$$

$$22.5 \ \text{g} \cdot \frac{1 \ \text{L}}{203.8 \ \text{g}} = 0.11 \ \text{L} \cdot \frac{1000 \ \text{mL}}{1 \ \text{L}} = 110 \ \text{mL}$$

21.

$$CuSO_4 + Fe \rightarrow FeSO_4 + Cu$$

$$7.7 \ \text{g} \ Cu \cdot \frac{\text{mol}}{63.546 \ \text{g}} = 0.1212 \ \text{mol}$$

$$0.1212 \ \text{mol} \cdot \frac{1 \ \text{mol} \ CuSO_4}{1 \ \text{mol} \ Cu} = 0.1212 \ \text{mol} \ CuSO_4$$

$$54.0 \ \text{mL} \cdot \frac{1 \ \text{L}}{1000 \ \text{mL}} = 0.0540 \ \text{L}$$

$$\frac{0.1212 \ \text{mol}}{0.0540 \ \text{L}} = 2.2 \ M$$

---

22. a.

$$Na_2SO_4(aq) + Ba(NO_3)_2(aq) \rightarrow 2NaNO_3(aq) + BaSO_4(s)$$

---

22. b.

$$475 \text{ mL} \cdot \frac{1 \text{ L}}{1000 \text{ mL}} = 0.475 \text{ L}$$

$$0.475 \text{ L} \cdot \frac{0.60 \text{ mol}}{\text{L}} = 0.285 \text{ mol Na}_2\text{SO}_4$$

$$0.285 \text{ mol Na}_2\text{SO}_4 \cdot \frac{1 \text{ mol BaSO}_4}{1 \text{ mol Na}_2\text{SO}_4} = 0.285 \text{ mol BaSO}_4$$

$$0.285 \text{ mol BaSO}_4 \cdot \frac{233.4 \text{ g}}{\text{mol}} = 67 \text{ g}$$

23.

$$H_2SO_4 + 2NaOH \rightarrow Na_2SO_4 + 2H_2O$$

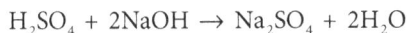

$$655 \text{ mL} \cdot \frac{1 \text{ L}}{1000 \text{ mL}} = 0.655 \text{ L}$$

$$0.655 \text{ L} \cdot \frac{6.00 \text{ mol}}{\text{L}} = 3.930 \text{ mol NaOH}$$

$$3.930 \text{ mol NaOH} \cdot \frac{1 \text{ mol H}_2\text{SO}_4}{2 \text{ mol NaOH}} = 1.965 \text{ mol H}_2\text{SO}_4$$

$$1.965 \text{ mol H}_2\text{SO}_4 \cdot \frac{\text{L}}{12.0 \text{ mol}} = 0.164 \text{ L} \cdot \frac{1000 \text{ mL}}{1 \text{ L}} = 164 \text{ mL}$$

24.

$$Fe + 2HCl \rightarrow FeCl_2 + H_2$$

$$2.00 \times 10^3 \text{ L} \cdot \frac{2.0 \text{ mol}}{\text{L}} = 4.00 \times 10^3 \text{ mol HCl}$$

$$4.00 \times 10^3 \text{ mol HCl} \cdot \frac{1 \text{ mol H}_2}{2 \text{ mol HCl}} = 2.00 \times 10^3 \text{ mol H}_2$$

$$P = 100 \text{ kPa}$$

$$T = 273.15 \text{ K}$$

$$R = 8.314 \frac{\text{L} \cdot \text{kPa}}{\text{mol} \cdot \text{K}}$$

$$PV = nRT \rightarrow V = \frac{nRT}{P} = \frac{2.00 \times 10^3 \text{ mol} \cdot 8.314 \frac{\text{L} \cdot \text{kPa}}{\text{mol} \cdot \text{K}} \cdot 273.15 \text{ K}}{100 \text{ kPa}} = 45,000 \text{ L}$$

---

25. a.

$$10.0 \frac{\text{mol}}{\text{kg}} \cdot 1.00 \text{ kg} = 10.0 \text{ mol HCl}$$

$$10.0 \text{ mol HCl} \cdot \frac{36.46 \text{ g}}{\text{mol}} = 365 \text{ g}$$

---

25. b.

$$1.75 \text{ L} \cdot \frac{1000 \text{ mL}}{1 \text{ L}} \cdot 0.998 \frac{\text{g}}{\text{mL}} = 1747 \text{ g} \cdot \frac{1 \text{ kg}}{1000 \text{ g}} = 1.747 \text{ kg}$$

$$5.25 \frac{\text{mol}}{\text{kg}} \cdot 1.747 \text{ kg} = 9.172 \text{ mol NaOH}$$

$$9.172 \text{ mol NaOH} \cdot \frac{40.00 \text{ g}}{\text{mol}} = 367 \text{ g}$$

26.

$$5.00 \times 10^2 \text{ mL} \cdot 0.998 \frac{\text{g}}{\text{mL}} = 499.0 \text{ g} \cdot \frac{1 \text{ kg}}{1000 \text{ g}} = 0.4990 \text{ kg}$$

$$225 \text{ g} \cdot \frac{\text{mol}}{342.3 \text{ g}} = 0.6573 \text{ mol}$$

$$\frac{0.6573 \text{ mol } C_{12}H_{22}O_{11}}{0.4990 \text{ kg H}_2O} = 1.32 \ m$$

27.

$$235 \text{ g} \cdot \frac{\text{mol}}{164.1 \text{ g}} = 1.432 \text{ mol}$$

$$\frac{1.432 \text{ mol Ca}(NO_3)_2}{x \text{ kg H}_2O} = 2.0 \frac{\text{mol}}{\text{kg}}$$

$$x = \frac{1.432 \text{ mol Ca}(NO_3)_2}{2.0 \frac{\text{mol}}{\text{kg}}} = 0.7161 \text{ kg H}_2O \cdot \frac{1000 \text{ g}}{\text{kg}} = 716.1 \text{ g}$$

$$716.1 \text{ g} \cdot \frac{\text{mL}}{0.998 \text{ g}} = 718 \text{ mL}$$

Rounding this value to 2 sig digs gives 720 mL.

---

30. c.

$$MgCl_2 + Pb(NO_3)_2 \rightarrow Mg(NO_3)_2 + PbCl_2$$

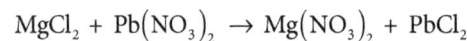

$$40.0 \text{ mL} \cdot \frac{1 \text{ L}}{1000 \text{ mL}} \cdot \frac{0.50 \text{ mol}}{\text{L}} = 0.0200 \text{ mol MgCl}_2$$

$$0.0200 \text{ mol MgCl}_2 \cdot \frac{1 \text{ PbCl}_2}{1 \text{ mol MgCl}_2} = 0.0200 \text{ mol PbCl}_2$$

$$0.0200 \text{ mol PbCl}_2 \cdot \frac{278.1 \text{ g}}{\text{mol}} = 5.6 \text{ g}$$

33.

$T_b = 99.974°C$

$K_b = 0.513 \dfrac{°C}{m}$

$\text{solute mass} = 0.4000 \text{ kg} \cdot \dfrac{1000 \text{ g}}{1 \text{ kg}} = 400.0 \text{ g } C_{12}H_{22}O_{11}$

$400.0 \text{ g } C_{12}H_{22}O_{11} \cdot \dfrac{\text{mol}}{342.30 \text{ g}} = 1.1686 \text{ mol}$

$1.100 \text{ L} \cdot \dfrac{1000 \text{ mL}}{1 \text{ L}} \cdot \dfrac{0.998 \text{ g}}{\text{mL}} \cdot \dfrac{1 \text{ kg}}{1000 \text{ g}} = 1.098 \text{ kg } H_2O$

$\text{molality} = \dfrac{1.1686 \text{ mol}}{1.098 \text{ kg}} = 1.0644 \ m$

$\Delta T_b = K_b m = 0.513 \dfrac{°C}{m} \cdot 1.0644 \ m = 0.546°C$

$T_b + \Delta T_b = 99.974°C + 0.546°C = 100.520°C$

34.

$T_b = 99.974°C$

$K_b = 0.513 \dfrac{°C}{m}$

First concentration:

$\Delta T_b = 101.75°C - 99.974°C = 1.776°C$

$\Delta T_b = K_b m \rightarrow m = \dfrac{\Delta T_b}{K_b} = \dfrac{1.776°C}{0.513 \dfrac{°C}{m}} = 3.46 \ m$

Second concentration:

$\Delta T_b = 100.99°C - 99.974°C = 1.016°C$

$\Delta T_b = K_b m \rightarrow m = \dfrac{\Delta T_b}{K_b} = \dfrac{1.016°C}{0.513 \dfrac{°C}{m}} = 1.98 \ m$

Third concentration:

$\Delta T_b = 103.17°C - 99.974°C = 3.196°C$

$\Delta T_b = K_b m \rightarrow m = \dfrac{\Delta T_b}{K_b} = \dfrac{3.196°C}{0.513 \dfrac{°C}{m}} = 6.23 \ m$

35.

$\Delta T_f = 18°C$

$K_f = 1.86 \dfrac{°C}{m}$

$\Delta T_f = K_f m \rightarrow m = \dfrac{\Delta T_f}{K_f} = \dfrac{18°C}{1.86 \dfrac{°C}{m}} = 9.7\ m$

36.

$\Delta T_f = 2.93°C$

$K_f = 1.86 \dfrac{°C}{m}$

mass of solvent $= 775\ g \cdot \dfrac{1\ kg}{1000\ g} = 0.775\ kg$

$\Delta T_f = K_f m \rightarrow m = \dfrac{\Delta T_f}{K_f} = \dfrac{2.93°C}{1.86 \dfrac{°C}{m}} = 1.575\ m$

$1.575\ m = \dfrac{\text{moles } CO(NH_2)_2}{0.775\ kg} \rightarrow \text{moles } CO(NH_2)_2 = 1.575\ m \cdot 0.775\ kg = 1.220\ mol$

$1.220\ mol\ CO(NH_2)_2 \cdot \dfrac{60.055\ g}{mol} = 73.3\ g$

37.

First, determine the mass of benzene:

s.g. $= 0.878$

$V = 2.50$ L

$$\rho_{water} = 0.998 \ \frac{g}{mL}$$

$$0.878 \cdot 0.998 \ \frac{g}{mL} = 0.8762 \ \frac{g}{mL} \cdot \frac{1000 \ mL}{1 \ L} = 876.2 \ \frac{g}{L}$$

$$876.2 \ \frac{g}{L} \cdot 2.50 \ L = 2191 \ g \cdot \frac{1 \ kg}{1000 \ g} = 2.191 \ kg \ C_6H_6$$

Next, determine the molality of the solution:

$$\Delta T_b = 85.34°C - 80.1°C = 5.24°C$$

$$K_b = 2.64 \ \frac{°C}{m}$$

$$\Delta T_b = K_b m \rightarrow m = \frac{\Delta T_b}{K_b} = \frac{5.24°C}{2.64 \ \frac{°C}{m}} = 1.98 \ m$$

Finally, put these together to determine the amount of CCl_4:

$$1.98 \ m = \frac{moles \ CCl_4}{2.191 \ kg \ C_6H_6} \rightarrow moles \ CCl_4 = 1.98 \ m \cdot 2.191 \ kg = 4.338 \ mol$$

$$4.338 \ mol \ CCl_4 \cdot \frac{153.8 \ g}{mol} = 667 \ g$$

This value must be rounded to 2 sig digs to give 670 g. This is because the calculation of ΔT_b using the addition rule gives 5.2°C (the 5.24°C has an extra digit), and this 2-digit value limits all subsequent results to 2 digits.

38.

Each mole of LiCl produces two moles of ions.

$$K_f = 1.86 \; \frac{°C}{m}$$

$$T_f = 0.00°C$$

$$575 \text{ g H}_2\text{O} \cdot \frac{1 \text{ kg}}{1000 \text{ g}} = 0.575 \text{ kg H}_2\text{O}$$

$$155 \text{ g} \cdot \frac{\text{mol}}{42.39 \text{ g}} = 3.656 \text{ mol LiCl}$$

$$3.656 \text{ mol LiCl} \cdot \frac{2 \text{ mol ions}}{1 \text{ mol LiCl}} = 7.312 \text{ mol ions}$$

$$m = \frac{7.312 \text{ mol ions}}{0.575 \text{ kg H}_2\text{O}} = 12.72 \; m$$

$$\Delta T_f = K_f m = 1.86 \; \frac{°C}{m} \cdot 12.72 \; m = 23.7°C$$

$$0.00°C - 23.7°C = -23.7°C$$

39.

$$K_f = 1.959 \; \frac{°C}{m}$$

$$T_f = -114.4°C$$

$$456 \text{ g C}_2\text{H}_5\text{OH} \cdot \frac{1 \text{ kg}}{1000 \text{ g}} = 0.456 \text{ kg C}_2\text{H}_5\text{OH}$$

$$75.0 \text{ g} \cdot \frac{\text{mol}}{74.55 \text{ g}} = 1.006 \text{ mol KCl}$$

$$1.006 \text{ mol KCl} \cdot \frac{2 \text{ mol ions}}{1 \text{ mol KCl}} = 2.012 \text{ mol ions}$$

$$m = \frac{2.012 \text{ mol ions}}{0.456 \text{ kg C}_2\text{H}_5\text{OH}} = 4.412 \; m$$

$$\Delta T_f = K_f m = 1.959 \; \frac{°C}{m} \cdot 4.412 \; m = 8.644°C$$

$$-114.4°C - 8.644°C = -123.0°C$$

40.

$$K_b = 3.22 \; \frac{°C}{m}$$

$$m = 0.040 \; m$$

$$0.040 \; m \; \text{KI} \cdot \frac{2 \; \text{mol ions}}{1 \; \text{mol KI}} = 0.080 \; m \; \text{KI ions}$$

$$\Delta T_b = K_b m = 3.22 \; \frac{°C}{m} \cdot 0.080 \; m = +0.26°C \; \text{(for KI)}$$

$$0.040 \; m \; \text{MgCl}_2 \cdot \frac{3 \; \text{mol ions}}{1 \; \text{mol KI}} = 0.120 \; m \; \text{MgCl}_2 \; \text{ions}$$

$$\Delta T_b = K_b m = 3.22 \; \frac{°C}{m} \cdot 0.120 \; m = +0.39°C \; \text{(for MgCl}_2)$$

$$0.040 \; m \; \text{CaSO}_4 \cdot \frac{2 \; \text{mol ions}}{1 \; \text{mol KI}} = 0.080 \; m \; \text{CaSO}_4 \; \text{ions}$$

$$\Delta T_b = K_b m = 3.22 \; \frac{°C}{m} \cdot 0.080 \; m = +0.26°C \; \text{(for CaSO}_4; \; \text{high ionic charges will likely}$$

make the actual change less than this prediction)

41.

$$T_f = 0.00°C$$

$$K_f = 1.86 \; \frac{°C}{m}$$

$$m = 0.020 \; m$$

For Na_3PO_4:

$$0.020 \; m \cdot \frac{4 \; \text{mol ions}}{1 \; \text{mol KI}} = 0.080 \; m \; \text{ions}$$

$$\Delta T_f = K_f m = 1.86 \; \frac{°C}{m} \cdot 0.080 \; m = 0.15°C$$

$$0.00 - 0.15°C = -0.15°C$$

For KCl:

$$0.020 \; m \cdot \frac{2 \; \text{mol ions}}{1 \; \text{mol KI}} = 0.040 \; m \; \text{ions}$$

$$\Delta T_f = K_f m = 1.86 \; \frac{°C}{m} \cdot 0.040 \; m = 0.07°C$$

$$0.00 - 0.07°C = -0.07°C$$

For $C_6H_{12}O_6$:

$$0.020 \; m \cdot \frac{1 \; \text{mol ions}}{1 \; \text{mol KI}} = 0.020 \; m \; \text{ions}$$

$$\Delta T_f = K_f m = 1.86 \; \frac{°C}{m} \cdot 0.020 \; m = 0.04°C$$

$$0.00 - 0.04°C = -0.04°C$$

For $CaCl_2$:

$$0.020 \; m \cdot \frac{3 \; \text{mol ions}}{1 \; \text{mol KI}} = 0.060 \; m \; \text{ions}$$

$$\Delta T_f = K_f m = 1.86 \; \frac{°C}{m} \cdot 0.060 \; m = 0.11°C$$

$$0.00 - 0.11°C = -0.11°C$$

42.

$m = 2.00\ m$

$K_f = 1.86\ \dfrac{°C}{m}$

$2.00\ m \cdot \dfrac{2\ \text{mol ions}}{1\ \text{mol HNO}_3} = 4.00\ m\ \text{ions}$

$\Delta T_f = K_f m = 1.86\ \dfrac{°C}{m} \cdot 4.00\ m = -7.44°C$

43.

$m = 1.30\ g$

$V = 2.24\ L$

$T = 6.85°C \rightarrow 6.85°C + 273.15 = 280.00\ K$

$P = 1.14\ atm$

$R = 0.08206\ \dfrac{L \cdot atm}{mol \cdot K}$

$PV = nRT \rightarrow n = \dfrac{PV}{RT} = \dfrac{1.14\ atm \cdot 2.24\ L}{0.08206\ \dfrac{L \cdot atm}{mol \cdot K} \cdot 280.00\ K} = 0.1111\ mol$

$M = \dfrac{1.30\ g}{0.1111\ mol} = 11.7\ \dfrac{g}{mol}$

44.

$CH_4(g) + 2O_2(g) \rightarrow CO_2(g) + 2H_2O(g)$

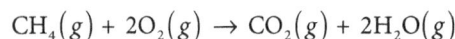

$17.5\ L\ CH_4 \cdot \dfrac{1\ \text{mol CO}_2}{1\ \text{mol CH}_4} = 17.5\ L\ CO_2$

$17.5\ L\ CH_4 \cdot \dfrac{2\ \text{mol H}_2O}{1\ \text{mol CH}_4} = 35.0\ L\ H_2O$

46.

Using F_w to indicate the weight,

$F_w = 12,500\ lb \cdot \dfrac{4.448\ N}{1\ lb} = 55,600\ N$

$l = 10.0\ in \cdot \dfrac{2.54\ cm}{in} \cdot \dfrac{1\ m}{100\ cm} = 0.254\ m$

$w = 10.0\ in \cdot \dfrac{2.54\ cm}{in} \cdot \dfrac{1\ m}{100\ cm} = 0.254\ m$

$A = l \cdot w = 0.254\ m \cdot 0.254\ m = 0.06452\ m^2$

$P = \dfrac{F}{A} = \dfrac{55,600\ N}{0.06452\ m^2} = 861,750\ Pa \cdot \dfrac{1\ atm}{101,325\ Pa} = 8.50\ atm$

47.

$$100.0 \text{ kg} \cdot \frac{1000 \text{ g}}{1 \text{ kg}} \cdot \frac{\text{mol}}{12.011 \text{ g}} = 8325.7 \text{ mol C}$$

$$8325.7 \text{ mol C} \cdot \frac{1 \text{ mol CH}_4}{2 \text{ mol C}} = 4162.9 \text{ mol CH}_4$$

$$4162.9 \text{ mol CH}_4 \cdot \frac{16.043 \text{ g}}{\text{mol}} = 66,780 \text{ g}$$

$$66,780 \text{ g} \cdot \frac{1 \text{ kg}}{1000 \text{ g}} \cdot 0.82 = 55 \text{ kg}$$

48.

$$125.0 \text{ mol C}_7\text{H}_6\text{O}_3 \cdot \frac{1 \text{ mol C}_9\text{H}_8\text{O}_4}{1 \text{ mol C}_7\text{H}_6\text{O}_3} = 125.0 \text{ mol C}_9\text{H}_8\text{O}_4$$

$$125.0 \text{ mol C}_9\text{H}_8\text{O}_4 \cdot \frac{180.16 \text{ g}}{\text{mol}} = 22,520.0 \text{ g}$$

$$22,520.0 \text{ g} \cdot \frac{1 \text{ kg}}{1000 \text{ g}} = 22.52 \text{ kg}$$

49.

$$\frac{7 \cdot 12.011}{138.123} = 0.60871 \rightarrow 60.871\% \text{ C}$$

$$\frac{6 \cdot 1.0079}{138.123} = 0.043783 \rightarrow 4.3783\% \text{ H}$$

$$\frac{3 \cdot 15.9994}{138.123} = 0.347503 \rightarrow 34.7503\% \text{ O}$$

Chapter 11

Note: In later printings, question 23 was renumbered to be question 26 to be included in Section 11.3. Accordingly, items below formerly numbeed as 24–26 have been renumbered 23–25.

23. a.

$$2HNO_3 + Na_2CO_3 \rightarrow CO_2 + 2NaNO_3 + H_2O$$

$$196 \text{ g Na}_2CO_3 \cdot \frac{\text{mol}}{105.99 \text{ g}} = 1.849 \text{ mol Na}_2CO_3$$

$$1.849 \text{ mol Na}_2CO_3 \cdot \frac{1 \text{ mol CO}_2}{1 \text{ mol Na}_2CO_3} = 1.849 \text{ mol CO}_2$$

$$1.849 \text{ mol CO}_2 \cdot \frac{44.01 \text{ g}}{\text{mol}} = 81.4 \text{ g CO}_2$$

$$P = 100 \text{ kPa}$$

$$T = 273.15 \text{ K}$$

$$R = 8.314 \frac{\text{L} \cdot \text{kPa}}{\text{mol} \cdot \text{K}}$$

$$PV = nRT \rightarrow V = \frac{nRT}{P} = \frac{1.849 \text{ mol} \cdot 8.314 \dfrac{\text{L} \cdot \text{kPa}}{\text{mol} \cdot \text{K}} \cdot 273.15 \text{ K}}{100 \text{ kPa}} = 42.0 \text{ L CO}_2$$

---

23. b.

$$1.849 \text{ mol Na}_2CO_3 \cdot \frac{2 \text{ mol HNO}_3}{1 \text{ mol Na}_2CO_3} = 3.698 \text{ mol HNO}_3$$

$$3.698 \text{ mol HNO}_3 \cdot \frac{\text{L}}{6.00 \text{ mol}} = 0.616 \text{ L} \cdot \frac{1000 \text{ mL}}{1 \text{ L}} = 616 \text{ mL}$$

---

24. a.

$$Cu + H_2SO_4 \rightarrow CuSO_4 + H_2$$

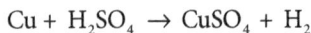

$$125 \text{ mL} \cdot \frac{1 \text{ L}}{1000 \text{ mL}} = 0.125 \text{ L}$$

$$0.125 \text{ L} \cdot \frac{8.0 \text{ mol}}{\text{L}} = 1.000 \text{ mol H}_2SO_4$$

$$1.000 \text{ mol H}_2SO_4 \cdot \frac{1 \text{ mol CuSO}_4}{1 \text{ mol H}_2SO_4} = 1.000 \text{ mol CuSO}_4$$

$$1.000 \text{ mol CuSO}_4 \cdot \frac{159.61 \text{ g}}{\text{mol}} = 160 \text{ g}$$

---

2. b.

$$1.000 \text{ mol } H_2SO_4 \cdot \frac{1 \text{ mol } H_2}{1 \text{ mol } H_2SO_4} = 1.000 \text{ mol } H_2$$

$P = 100 \text{ kPa}$

$T = 273.15 \text{ K}$

$$R = 8.314 \frac{L \cdot kPa}{mol \cdot K}$$

$$PV = nRT \rightarrow V = \frac{nRT}{P} = \frac{1.000 \text{ mol} \cdot 8.314 \frac{L \cdot kPa}{mol \cdot K} \cdot 273.15 \text{ K}}{100 \text{ kPa}} = 22.7 \text{ L}$$

---

25. a.

$$CaCO_3 + 2HCl \rightarrow CaCl_2 + CO_2 + H_2O$$

$$V = 2.5 \times 10^4 \text{ mL} \cdot \frac{1 \text{ L}}{1000 \text{ mL}} = 25 \text{ L } CO_2$$

$P = 100 \text{ kPa}$

$T = 273.15$

$$R = 8.314 \frac{L \cdot kPa}{mol \cdot K}$$

$$PV = nRT \rightarrow n = \frac{PV}{RT} = \frac{100 \text{ kPa} \cdot 25 \text{ L}}{8.314 \frac{L \cdot kPa}{mol \cdot K} \cdot 273.15} = 1.10 \text{ mol } CO_2$$

$$1.10 \text{ mol } CO_2 \cdot \frac{1 \text{ mol } CaCO_3}{1 \text{ mol } CO_2} = 1.10 \text{ mol } CaCO_3$$

$$1.10 \text{ mol } CaCO_3 \cdot \frac{100.1 \text{ g}}{mol} = 110 \text{ g}$$

---

25. b

$$1.10 \text{ mol } CO_2 \cdot \frac{2 \text{ mol } HCl}{1 \text{ mol } CO_2} = 2.20 \text{ mol } HCl$$

$$2.20 \text{ mol } HCl \cdot \frac{L}{2.00 \text{ mol}} = 1.1 \text{ L}$$

---

31. a

$$[H_3O^+] = 1.56 \times 10^{-3} \ M$$

$$[H_3O^+] \cdot [OH^-] = 1.0 \times 10^{-14} \ M^2$$

$$[OH^-] = \frac{1.0 \times 10^{-14} \ M^2}{[H_3O^+]} = \frac{1.0 \times 10^{-14} \ M^2}{1.56 \times 10^{-3} \ M} = 6.4 \times 10^{-12} \ M$$

---

31. b.

Each mole of $Sr(OH)_2$ contributes two moles of OH^- ions.

$\left[OH^-\right]=2\cdot0.0110\ M=0.0220\ M$

$\left[H_3O^+\right]\cdot\left[OH^-\right]=1.0\times10^{-14}\ M^2$

$\left[H_3O^+\right]=\dfrac{1.0\times10^{-14}\ M^2}{\left[OH^-\right]}=\dfrac{1.0\times10^{-14}\ M^2}{0.0220\ M}=4.5\times10^{-13}\ M$

31. c.

$\left[OH^-\right]=2.50\ M$

$\left[H_3O^+\right]\cdot\left[OH^-\right]=1.0\times10^{-14}\ M^2$

$\left[H_3O^+\right]=\dfrac{1.0\times10^{-14}\ M^2}{\left[OH^-\right]}=\dfrac{1.0\times10^{-14}\ M^2}{2.50\ M}=4.0\times10^{-15}\ M$

31. d.

$\left[H_3O^+\right]=8.911\times10^{-5}\ M$

$\left[H_3O^+\right]\cdot\left[OH^-\right]=1.0\times10^{-14}\ M^2$

$\left[OH^-\right]=\dfrac{1.0\times10^{-14}\ M^2}{\left[H_3O^+\right]}=\dfrac{1.0\times10^{-14}\ M^2}{8.911\times10^{-5}\ M}=1.1\times10^{-10}\ M$

32.

In each case, the number of sig digs in the result is determined by two principles: 1) the number of sig digs in a number you are taking a logarithm of must equal the number of digits to the right of the decimal in your logarithm; 2) after taking a logarithm to obtain pH or pOH, the other one is obtained from an equation using either the addition rule and the sig digs are determined accordingly.

32. a.

$pOH=-\log\left[OH^-\right]=-\log0.0250=1.602$

$pH+pOH=14.0$

$pH=14.0-pOH=14.0-1.602=12.4$

32. b.

$pH=-\log\left[H_3O^+\right]=-\log6.0=-0.78$

$pH+pOH=14.0$

$pOH=14.0-pH=14.0-(-0.78)=14.8$

32. c.

$$pH = -\log\left[H_3O^+\right] = -\log 12 = -1.08$$
$$pH + pOH = 14.0$$
$$pOH = 14.0 - pH = 14.0 - (-1.08) = 15.1$$

---

32. d.

$$\left[OH^-\right] = 2 \cdot 2.03 \times 10^{-2}\ M = 4.06 \times 10^{-2}\ M$$
$$pOH = -\log\left[OH^-\right] = -\log(4.06 \times 10^{-2}) = 1.391$$
$$pH + pOH = 14.0$$
$$pH = 14.0 - pOH = 14.0 - (1.391) = 12.6$$

33.

When computing with antilogs, the number of sig digs in the antilog must equal the number of digits to the right of the decimal in the value you started with. After using an antilog obtain $[H_3O^+]$ or $[OH^-]$, the other concentration is obtained from an equation using the multiplication rule and the sig digs are determined accordingly.

33. a.

$$pH = 1.00$$
$$\left[H_3O^+\right] = 10^{-pH} = 10^{-1.00} = 0.10\ M$$
$$\left[H_3O^+\right] \cdot \left[OH^-\right] = 1.0 \times 10^{-14}\ M^2$$
$$\left[OH^-\right] = \frac{1.0 \times 10^{-14}\ M^2}{\left[H_3O^+\right]} = \frac{1.0 \times 10^{-14}\ M^2}{0.10\ M} = 1.0 \times 10^{-13}\ M$$

---

33. b.

$$pH = 3.5$$
$$\left[H_3O^+\right] = 10^{-pH} = 10^{-3.5} = 0.0003\ M$$
$$\left[H_3O^+\right] \cdot \left[OH^-\right] = 1.0 \times 10^{-14}\ M^2$$
$$\left[OH^-\right] = \frac{1.0 \times 10^{-14}\ M^2}{\left[H_3O^+\right]} = \frac{1.0 \times 10^{-14}\ M^2}{0.0003\ M} = 3 \times 10^{-11}\ M$$

---

33. c.

$$pOH = 10.0$$
$$\left[OH^-\right] = 10^{-pOH} = 10^{-10.0} = 1 \times 10^{-10}\ M$$
$$\left[H_3O^+\right] \cdot \left[OH^-\right] = 1.0 \times 10^{-14}\ M^2$$
$$\left[H_3O^+\right] = \frac{1.0 \times 10^{-14}\ M^2}{\left[OH^-\right]} = \frac{1.0 \times 10^{-14}\ M^2}{1 \times 10^{-10}\ M} = 1 \times 10^{-4}\ M$$

33. d.

pH = 11.88

$\left[H_3O^+\right] = 10^{-pH} = 10^{-11.88} = 1.3 \times 10^{-12} \ M$

$\left[H_3O^+\right] \cdot \left[OH^-\right] = 1.0 \times 10^{-14} \ M^2$

$\left[OH^-\right] = \dfrac{1.0 \times 10^{-14} \ M^2}{\left[H_3O^+\right]} = \dfrac{1.0 \times 10^{-14} \ M^2}{1.3 \times 10^{-12} \ M} = 0.0077 \ M$

---

33. e.

pOH = 2.1

$\left[OH^-\right] = 10^{-pOH} = 10^{-2.1} = 0.008 \ M$

$\left[H_3O^+\right] \cdot \left[OH^-\right] = 1.0 \times 10^{-14} \ M^2$

$\left[H_3O^+\right] = \dfrac{1.0 \times 10^{-14} \ M^2}{\left[OH^-\right]} = \dfrac{1.0 \times 10^{-14} \ M^2}{0.008 \ M} = 1 \times 10^{-12} \ M$

---

33. f.

pOH = 7.0

$\left[OH^-\right] = 10^{-pOH} = 10^{-7.0} = 1 \times 10^{-7} \ M$

$\left[H_3O^+\right] \cdot \left[OH^-\right] = 1.0 \times 10^{-14} \ M^2$

$\left[H_3O^+\right] = \dfrac{1.0 \times 10^{-14} \ M^2}{\left[OH^-\right]} = \dfrac{1.0 \times 10^{-14} \ M^2}{1 \times 10^{-7} \ M} = 1 \times 10^{-7} \ M$

---

33. g.

pOH = 4.00

$\left[OH^-\right] = 10^{-pOH} = 10^{-4.00} = 1.0 \times 10^{-4} \ M$

$\left[H_3O^+\right] \cdot \left[OH^-\right] = 1.0 \times 10^{-14} \ M^2$

$\left[H_3O^+\right] = \dfrac{1.0 \times 10^{-14} \ M^2}{\left[OH^-\right]} = \dfrac{1.0 \times 10^{-14} \ M^2}{1.0 \times 10^{-4} \ M} = 1.0 \times 10^{-10} \ M$

---

33. h.

pH = 14.0

$\left[H_3O^+\right] = 10^{-pH} = 10^{-14.0} = 1 \times 10^{-14} \ M$

$\left[H_3O^+\right] \cdot \left[OH^-\right] = 1.0 \times 10^{-14} \ M^2$

$\left[OH^-\right] = \dfrac{1.0 \times 10^{-14} \ M^2}{\left[H_3O^+\right]} = \dfrac{1.0 \times 10^{-14} \ M^2}{1 \times 10^{-14} \ M} = 1 \ M$

---

34. a.

pH $= 5.6$

$pH + pOH = 14.0$

$pOH = 14.0 - pH = 14.0 - 5.6 = 8.4$

$\left[H_3O^+\right] = 10^{-pH} = 10^{-5.6} = 3\times10^{-6}\ M$

$\left[OH^-\right] = 10^{-pOH} = 10^{-8.4} = 4\times10^{-9}\ M$

Alternately, this calculation for $\left[OH^-\right]$ leads to a slightly different result due to limits on precision:

$\left[H_3O^+\right]\cdot\left[OH^-\right] = 1.0\times10^{-14}\ M^2$

$\left[OH^-\right] = \dfrac{1.0\times10^{-14}\ M^2}{\left[H_3O^+\right]} = \dfrac{1.0\times10^{-14}\ M^2}{3\times10^{-6}\ M} = 3\times10^{-9}\ M$

---

34. b.

$\left[OH^-\right] = 5.5\times10^{-10}\ M$

$\left[H_3O^+\right]\cdot\left[OH^-\right] = 1.0\times10^{-14}\ M^2$

$\left[H_3O^+\right] = \dfrac{1.0\times10^{-14}\ M^2}{\left[OH^-\right]} = \dfrac{1.0\times10^{-14}\ M^2}{5.5\times10^{-10}\ M} = 1.8\times10^{-5}\ M$

$pOH = -\log\left[OH^-\right] = -\log 5.5\times10^{-10} = 9.26$

$pH + pOH = 14.0$

$pH = 14.0 - pOH = 14.0 - 9.26 = 4.7$

---

34. c.

$\left[H_3O^+\right] = 1.0\times10^{-9}\ M$

$\left[H_3O^+\right]\cdot\left[OH^-\right] = 1.0\times10^{-14}\ M^2$

$\left[OH^-\right] = \dfrac{1.0\times10^{-14}\ M^2}{\left[H_3O^+\right]} = \dfrac{1.0\times10^{-14}\ M^2}{1.0\times10^{-9}\ M} = 1.0\times10^{-5}\ M$

$pH = -\log\left[H_3O^+\right] = -\log\left(1.0\times10^{-9}\right) = 9.00$

For highest precision, calculate pOH this way:

$pOH = -\log\left[OH^-\right] = -\log\left(1.0\times10^{-5}\right) = 5.00$

---

34. d.

pOH $=11.5$

pH $+$ pOH $=14.0$

pH $=14.0-$ pOH $=14.0-11.5=2.5$

$\left[OH^-\right]=10^{-pOH}=10^{-11.5}=3\times10^{-12}\ M$

$\left[H_3O^+\right]=10^{-pH}=10^{-2.5}=0.003\ M$

Alternately:

$\left[H_3O^+\right]\cdot\left[OH^-\right]=1.0\times10^{-14}\ M^2$

$\left[H_3O^+\right]=\dfrac{1.0\times10^{-14}\ M^2}{\left[OH^-\right]}=\dfrac{1.0\times10^{-14}\ M^2}{3\times10^{-12}\ M}=0.003\ M$

---

34. e.

$\left[H_3O^+\right]=4.5\times10^{-3}\ M$

$\left[H_3O^+\right]\cdot\left[OH^-\right]=1.0\times10^{-14}\ M^2$

$\left[OH^-\right]=\dfrac{1.0\times10^{-14}\ M^2}{\left[H_3O^+\right]}=\dfrac{1.0\times10^{-14}\ M^2}{4.5\times10^{-3}\ M}=2.2\times10^{-12}\ M$

$pH=-\log\left[H_3O^+\right]=-\log\left(4.5\times10^{-3}\right)=2.35$

For highest precision, calculate pOH this way:

$pOH=-\log\left[OH^-\right]=-\log\left(2.2\times10^{-12}\right)=11.66$

---

34. f.

pOH $=1.00$

pH $+$ pOH $=14.0$

pH $=14.0-$ pOH $=14.0-1.00=13.0$

$\left[OH^-\right]=10^{-pOH}=10^{-1.00}=0.10\ M$

For highest precision, calculate $\left[H_3O^+\right]$ this way:

$\left[H_3O^+\right]\cdot\left[OH^-\right]=1.0\times10^{-14}\ M^2$

$\left[H_3O^+\right]=\dfrac{1.0\times10^{-14}\ M^2}{\left[OH^-\right]}=\dfrac{1.0\times10^{-14}\ M^2}{0.10\ M}=1.0\times10^{-13}\ M$

---

34. g.

pOH = 7.22

pH + pOH = 14.0

pH = 14.0 − pOH = 14.0 − 7.22 = 6.8

$\left[OH^-\right] = 10^{-pOH} = 10^{-7.22} = 6.0 \times 10^{-8} \ M$

$\left[H_3O^+\right] = 10^{-pH} = 10^{-6.8} = 2 \times 10^{-7} \ M$

Alternately, this calculation for $\left[H_3O^+\right]$ allows an extra sig dig:

$\left[H_3O^+\right] \cdot \left[OH^-\right] = 1.0 \times 10^{-14} \ M^2$

$\left[H_3O^+\right] = \dfrac{1.0 \times 10^{-14} \ M^2}{\left[OH^-\right]} = \dfrac{1.0 \times 10^{-14} \ M^2}{6.0 \times 10^{-8} \ M} = 1.7 \times 10^{-7} \ M$

---

34. h.

pH = 13.50

pH + pOH = 14.0

pOH = 14.0 − pH = 14.0 − 13.50 = 0.5

$\left[H_3O^+\right] = 10^{-pH} = 10^{-13.50} = 3.2 \times 10^{-14} \ M$

$\left[OH^-\right] = 10^{-pOH} = 10^{-0.5} = 0.3 \ M$

Alternately, this calculation for $\left[OH^-\right]$ allows an extra digit of precision:

$\left[H_3O^+\right] \cdot \left[OH^-\right] = 1.0 \times 10^{-14} \ M^2$

$\left[OH^-\right] = \dfrac{1.0 \times 10^{-14} \ M^2}{\left[H_3O^+\right]} = \dfrac{1.0 \times 10^{-14} \ M^2}{3.2 \times 10^{-14} \ M} = 0.31 \ M$

---

34. i.

pH = 2.0

pH + pOH = 14.0

pOH = 14.0 − pH = 14.0 − 2.0 = 12.0

$\left[H_3O^+\right] = 10^{-pH} = 10^{-2.0} = 0.01 \ M$

$\left[OH^-\right] = 10^{-pOH} = 10^{-12.0} = 1 \times 10^{-12} \ M$

Alternately:

$\left[H_3O^+\right] \cdot \left[OH^-\right] = 1.0 \times 10^{-14} \ M^2$

$\left[OH^-\right] = \dfrac{1.0 \times 10^{-14} \ M^2}{\left[H_3O^+\right]} = \dfrac{1.0 \times 10^{-14} \ M^2}{0.01 \ M} = 1 \times 10^{-12} \ M$

---

34. j.

$$\left[OH^- \right] = 3.2 \times 10^{-2} \ M$$

$$\left[H_3O^+ \right] \cdot \left[OH^- \right] = 1.0 \times 10^{-14} \ M^2$$

$$\left[H_3O^+ \right] = \frac{1.0 \times 10^{-14} \ M^2}{\left[OH^- \right]} = \frac{1.0 \times 10^{-14} \ M^2}{3.2 \times 10^{-2} \ M} = 3.1 \times 10^{-13} \ M$$

$$pOH = -\log\left[OH^- \right] = -\log\left(3.2 \times 10^{-2} \ M \right) = 1.49$$

$$pH = -\log\left[H_3O^+ \right] = -\log\left(3.1 \times 10^{-13} \ M \right) = 12.51$$

---

35. a.

$$24.05 \ mL \cdot \frac{1 \ L}{1000 \ mL} \cdot 1.65 \times 10^{-3} \ M = 3.968 \times 10^{-5} \ mol \ NaOH$$

$$3.968 \times 10^{-5} \ mol \ NaOH \rightarrow 3.968 \times 10^{-5} \ mol \ HNO_3$$

$$M_{HNO_3} \cdot 18.55 \ mL \cdot \frac{1 \ L}{1000 \ mL} = 3.968 \times 10^{-5} \ mol \ HNO_3$$

$$M_{HNO_3} = \frac{3.968 \times 10^{-5} \ mol \ HNO_3}{0.01855 \ L} = 2.14 \times 10^{-3} \ M \ HNO_3$$

---

35. b.

$$21.00 \ mL \cdot \frac{1 \ L}{1000 \ mL} \cdot 1.500 \ M = 0.03150 \ mol \ HClO_3$$

$$0.03150 \ mol \ HClO_3 \rightarrow 0.03150 \ mol \ KOH$$

$$M_{KOH} \cdot 23.78 \ mL \cdot \frac{1 \ L}{1000 \ mL} = 0.03150 \ mol \ KOH$$

$$M_{KOH} = \frac{0.03150 \ mol \ KOH}{0.02378 \ L} = 1.325 \ M \ KOH$$

---

35. c.

$$20.87 \ mL \cdot \frac{1 \ L}{1000 \ mL} \cdot 4.077 \times 10^{-2} \ M = 8.509 \times 10^{-4} \ mol \ HClO_4$$

$Ca(OH)_2$ provides two moles of OH^- ions for each mole of $Ca(OH)_2$. So only half as many moles of $Ca(OH)_2$ are required to neutralize $HClO_4$.

$$\frac{8.509 \times 10^{-4} \ mol \ HClO_4}{2} \rightarrow 4.254 \times 10^{-4} \ mol \ Ca(OH)_2$$

$$M_{Ca(OH)_2} \cdot 22.94 \ mL \cdot \frac{1 \ L}{1000 \ mL} = 4.254 \times 10^{-4} \ mol \ Ca(OH)_2$$

$$M_{Ca(OH)_2} = \frac{4.254 \times 10^{-4} \ mol \ Ca(OH)_2}{0.02294 \ L} = 0.01855 \ M \ Ca(OH)_2$$

---

35. d.

$$20.05 \text{ mL} \cdot \frac{1 \text{ L}}{1000 \text{ mL}} \cdot 3.050 \, M = 0.06115 \text{ mol LiOH}$$

$$0.06115 \text{ mol LiOH} \rightarrow 0.06115 \text{ mol HCl}$$

$$M_{\text{HCl}} \cdot 22.00 \text{ mL} \cdot \frac{1 \text{ L}}{1000 \text{ mL}} = 0.06115 \text{ mol HCl}$$

$$M_{\text{HCl}} = \frac{0.06115 \text{ mol HCl}}{0.02200 \text{ L}} = 2.780 \, M \text{ HCl}$$

36.

$$23.09 \text{ mL} \cdot \frac{1 \text{ L}}{1000 \text{ mL}} \cdot 5.06 \times 10^{-3} \, M = 1.1684 \times 10^{-4} \text{ mol HCl}$$

$$1.1684 \times 10^{-4} \text{ mol HCl} \rightarrow 1.1684 \times 10^{-4} \text{ mol KOH}$$

$$M_{\text{KOH}} \cdot 20.07 \text{ mL} \cdot \frac{1 \text{ L}}{1000 \text{ mL}} = 1.1684 \times 10^{-4} \text{ mol KOH}$$

$$M_{\text{KOH}} = \frac{1.1684 \times 10^{-4} \text{ mol KOH}}{0.02007 \text{ L}} = 5.82 \times 10^{-3} \, M \text{ KOH}$$

37.

$$22.01 \text{ mL} \cdot \frac{1 \text{ L}}{1000 \text{ mL}} \cdot 0.0231 \, M = 5.084 \times 10^{-4} \text{ mol Ba(OH)}_2$$

$Ba(OH)_2$ provides two moles of OH^- ions for each mole of $Ba(OH)_2$. So twice as many moles of HCl will be neutralized.

$$2 \cdot 5.084 \times 10^{-4} \text{ mol Ba(OH)}_2 \rightarrow 1.017 \times 10^{-3} \text{ mol HCl}$$

$$M_{\text{HCl}} \cdot 27.0 \text{ mL} \cdot \frac{1 \text{ L}}{1000 \text{ mL}} = 1.017 \times 10^{-3} \text{ mol HCl}$$

$$M_{\text{HCl}} = \frac{1.017 \times 10^{-3} \text{ mol HCl}}{0.0270 \text{ L}} = 0.0377 \, M \text{ HCl}$$

38.

$$23.90 \text{ mL} \cdot \frac{1 \text{ L}}{1000 \text{ mL}} \cdot 2.5 \times 10^{-3} \, M = 5.98 \times 10^{-5} \text{ mol NaOH}$$

5.98×10^{-5} mol NaOH neutralizes 5.98×10^{-5} mol acid, so 5.98×10^{-5} mol acid ionized.

$$20.00 \text{ mL} \cdot \frac{1 \text{ L}}{1000 \text{ mL}} \cdot 0.04098 \, M = 8.196 \times 10^{-4} \text{ mol acid present}$$

$$\frac{5.98 \times 10^{-5} \text{ mol acid ionized}}{8.196 \times 10^{-4} \text{ mol acid present}} = 0.073 \rightarrow 7.3\% \text{ ionization}$$

39.

$$\frac{1.77 \text{ g}}{100 \text{ mL}} \cdot \frac{1000 \text{ mL}}{1 \text{ L}} \cdot \frac{\text{mol}}{121.64 \text{ g}} = 0.146 \, M \text{ (maximum concentration)}$$

40.

$$38.95 \text{ g} \cdot \frac{\text{mol}}{342.299 \text{ g}} = 0.11379 \text{ mol}$$

$$\frac{0.11379 \text{ mol}}{250.0 \text{ mL}} \cdot \frac{1000 \text{ mL}}{1 \text{ L}} = 0.4552 \text{ M}$$

41.

$$n = 37.05 \text{ g} \cdot \frac{\text{mol}}{18.015 \text{ g}} = 2.0566 \text{ mol}$$

$$C_{ice} = 0.0364 \ \frac{\text{kJ}}{\text{mol} \cdot \text{K}}$$

$$C_{water} = 0.0752 \ \frac{\text{kJ}}{\text{mol} \cdot \text{K}}$$

$$H_f = 6.01 \ \frac{\text{kJ}}{\text{mol}}$$

$$\Delta T_{ice} = 0.0°C - (-18.0°C) = 18.0°C = 18.0 \text{ K}$$

$$\Delta T_{water} = 20.0°C - 0.0°C = 20.0°C = 20.0 \text{ K}$$

To make the sig digs clear for the final addition step, we will list the results of the heat calculations without any extra sig digs.

To warm the ice to 0°C:

$$Q = Cn\Delta T = 0.0364 \ \frac{\text{kJ}}{\text{mol} \cdot \text{K}} \cdot 2.0566 \text{ mol} \cdot 18.0 \text{ K} = 1.35 \text{ kJ}$$

To melt the ice:

$$Q = nH_f = 2.0566 \text{ mol} \cdot 6.01 \ \frac{\text{kJ}}{\text{mol}} = 12.4 \text{ kJ}$$

To warm the water:

$$Q = Cn\Delta T = 0.0752 \ \frac{\text{kJ}}{\text{mol} \cdot \text{K}} \cdot 2.0566 \text{ mol} \cdot 20.0 \text{ K} = 3.09 \text{ kJ}$$

$$Q_{total} = 1.35 \text{ kJ} + 12.4 \text{ kJ} + 3.09 \text{ kJ} = 16.8 \text{ kJ}$$

42. a.

$$NaOH + HNO_3 \rightarrow NaNO_3 + H_2O$$

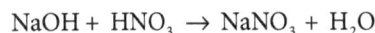

$$35.55 \text{ g NaOH} \cdot \frac{\text{mol}}{39.997 \text{ g}} = 0.88881 \text{ mol NaOH}$$

$$157.0 \text{ mL HNO}_3 \cdot \frac{1 \text{ L}}{1000 \text{ mL}} \cdot \frac{4.777 \text{ mol}}{\text{L}} = 0.74999 \text{ mol HNO}_3$$

The mole ratio is 1:1, so HNO_3 is the limiting reactant.

42. b.

$$0.74999 \text{ mol HNO}_3 \cdot \frac{1 \text{ mol NaNO}_3}{1 \text{ mol HNO}_3} = 0.7500 \text{ mol NaNO}_3$$

42. c.

$$0.74999 \text{ mol HNO}_3 \cdot \frac{1 \text{ mol H}_2\text{O}}{1 \text{ mol HNO}_3} = 0.7500 \text{ mol H}_2\text{O}$$

42. d.

$$0.8888 \text{ mol NaOH} - 0.7500 \text{ mol NaOH} = 0.1388 \text{ mol}$$

44. c.

$$25.50 \text{ mL} \cdot \frac{1 \text{ L}}{1000 \text{ mL}} \cdot \frac{0.050 \text{ mol}}{\text{L}} = 0.00128 \text{ mol Na}_2\text{CO}_3$$

$$0.00128 \text{ mol Na}_2\text{CO}_3 \cdot \frac{1 \text{ mol BaCO}_3}{1 \text{ mol Na}_2\text{CO}_3} = 0.00128 \text{ mol BaCO}_3$$

$$0.00128 \text{ mol BaCO}_3 \cdot \frac{197.336 \text{ g}}{\text{mol}} = 0.25 \text{ g}$$

45.

$$\text{Cu} + \text{H}_2\text{SO}_4 \rightarrow \text{CuSO}_4 + \text{H}_2$$

$$V = 109 \text{ mL H}_2 \cdot \frac{1 \text{ L}}{1000 \text{ mL}} = 0.109 \text{ L}$$

$$P_T = 765 \text{ Torr}$$

$$T = 25°\text{C} \rightarrow 25°\text{C} + 273.2 = 298.2 \text{ K}$$

$$R = 62.36 \frac{\text{L} \cdot \text{Torr}}{\text{mol} \cdot \text{K}}$$

At these conditions, the vapor pressure of water is 23.78 Torr.

$$P_T = P_{\text{H}_2} + P_{\text{H}_2\text{O}}$$

$$P_{\text{H}_2} = P_T - P_{\text{H}_2\text{O}} = 765 \text{ Torr} - 23.78 \text{ Torr} = 741.2 \text{ Torr}$$

$$PV = nRT \rightarrow n_{\text{H}_2} = \frac{P_{\text{H}_2}V}{RT} = \frac{741.2 \text{ Torr} \cdot 0.109 \text{ L}}{62.36 \dfrac{\text{L} \cdot \text{Torr}}{\text{mol} \cdot \text{K}} \cdot 298.2 \text{ K}} = 0.004345 \text{ mol H}_2$$

$$0.004345 \text{ mol H}_2 \cdot \frac{1 \text{ mol Cu}}{1 \text{ mol H}_2} = 0.004345 \text{ mol Cu}$$

$$0.004345 \text{ mol Cu} \cdot \frac{63.55 \text{ g}}{\text{mol}} = 0.276 \text{ g}$$

46. a.

$V = 2.500$ L

$P = 13.11$ atm

$T = 22°C \rightarrow 22°C + 273.2 = 295.2$ K

$R = 0.08206 \dfrac{\text{L} \cdot \text{Torr}}{\text{mol} \cdot \text{K}}$

$$PV = nRT \rightarrow n = \frac{PV}{RT} = \frac{13.11 \text{ atm} \cdot 2.500 \text{ L}}{0.08206 \dfrac{\text{L} \cdot \text{Torr}}{\text{mol} \cdot \text{K}} \cdot 295.2 \text{ K}} = 1.3530 \text{ mol}$$

Rounding to 4 sig digs we have 1.353 mol.

46. b.

$$1.3530 \text{ mol O}_2 \cdot \frac{31.9988 \text{ g}}{\text{mol}} = 43.29 \text{ g}$$

47. a.

$$\text{Fe}(s) + \text{CuSO}_4(aq) \rightarrow \text{FeSO}_4(aq) + \text{Cu}(s)$$

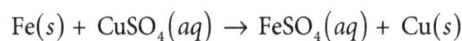

$$25.00 \text{ g Fe} \cdot \frac{\text{mol}}{55.847 \text{ g}} = 0.44765 \text{ mol Fe}$$

$$35.8 \text{ g CuSO}_4 \cdot \frac{\text{mol}}{159.610 \text{ g}} = 0.2243 \text{ mol CuSO}_4$$

Since the mole ratio is 1:1, CuSO_4 is the limiting reactant.

47. b.

$$0.2243 \text{ mol CuSO}_4 \cdot \frac{1 \text{ mol FeSO}_4}{1 \text{ mol CuSO}_4} = 0.2243 \text{ mol FeSO}_4$$

$$0.2243 \text{ mol FeSO}_4 \cdot \frac{151.911 \text{ g}}{\text{mol}} = 34.1 \text{ g}$$

47. c.

$0.4477 \text{ mol Fe} - 0.2243 \text{ mol Fe} = 0.2234 \text{ mol Fe}$

$$0.2234 \text{ mol Fe} \cdot \frac{55.847 \text{ g}}{\text{mol}} = 12.5 \text{ g}$$

48.

$m = 8.755$ g

$V = 2.00$ L

$P = 1.05 \text{ bar} \cdot \dfrac{100 \text{ kPa}}{1 \text{ bar}} = 105 \text{ kPa}$

$T = 30.0°\text{C} \rightarrow 30.0°\text{C} + 273.2 = 303.2$ K

$R = 8.314 \dfrac{\text{L} \cdot \text{kPa}}{\text{mol} \cdot \text{K}}$

$PV = nRT \rightarrow n = \dfrac{PV}{RT} = \dfrac{105 \text{ kPa} \cdot 2.00 \text{ L}}{8.314 \dfrac{\text{L} \cdot \text{kPa}}{\text{mol} \cdot \text{K}} \cdot 303.2} = 0.08331$ mol

$M = \dfrac{8.755 \text{ g}}{0.08331 \text{ mol}} = 105 \dfrac{\text{g}}{\text{mol}}$

49.

$\lambda = 0.22 \text{ nm} \cdot \dfrac{1 \text{ m}}{10^9 \text{ nm}} = 2.2 \times 10^{-10}$ m

$E = \dfrac{hv}{\lambda} = \dfrac{6.626 \times 10^{-34} \text{ J} \cdot \text{s} \cdot 2.9979 \times 10^8 \dfrac{\text{m}}{\text{s}}}{2.2 \times 10^{-10} \text{ m}} = 9.03 \times 10^{-16}$ J

$9.03 \times 10^{-16} \text{ J} \cdot \dfrac{1 \text{ eV}}{1.602 \times 10^{-19} \text{ J}} = 5600 \text{ eV}$

Chapter 12

6. a.

Add one H_2O to the left to balance the oxygen. Add $2H^+$ to the right to balance the hydrogen. Add $2OH^-$ to each side to "neutralize" the $2H^+$. Add two electrons to the right to balance the charges.

$$SO_3^{2-}(aq) + H_2O(l) \rightarrow SO_4^{2-}(aq) + 2H^+(aq)$$
$$SO_3^{2-}(aq) + H_2O(l) + 2OH^-(aq) \rightarrow SO_4^{2-}(aq) + 2H^+(aq) + 2OH^-(aq)$$
$$SO_3^{2-}(aq) + H_2O(l) + 2OH^-(aq) \rightarrow SO_4^{2-}(aq) + 2H^+(aq) + 2OH^-(aq) + 2e^-$$

On the right, combine the H^+ and OH^- ions to form $2H_2O$. Cancel out one H_2O from each side.

$$SO_3^{2-}(aq) + H_2O(l) + 2OH^-(aq) \rightarrow SO_4^{2-}(aq) + 2H_2O(l) + 2e^-$$
$$SO_3^{2-}(aq) + 2OH^-(aq) \rightarrow SO_4^{2-}(aq) + H_2O(l) + 2e^-$$

6. b.

Begin by placing a coefficient of 2 on the NH_3 to balance the nitrogen. Add $6H^+$ to the left to balance the hydrogen. Add $6OH^-$ to each side to neutralize the H^+. Combine the ions on the left to form $6H_2O$. Add six electrons to the left to balance the charges.

$$N_2(g) \rightarrow 2NH_3(g)$$
$$N_2(g) + 6H^+(aq) + 6OH^-(aq) \rightarrow 2NH_3(g) + 6OH^-(aq)$$
$$N_2(g) + 6H_2O(l) \rightarrow 2NH_3(g) + 6OH^-(aq)$$
$$N_2(g) + 6H_2O(l) + 6e^- \rightarrow 2NH_3(g) + 6OH^-(aq)$$

6. c.

$$N_2(g) \rightarrow 2NH_4^+(aq)$$
$$N_2(g) + 8H^+(aq) \rightarrow 2NH_4^+(aq)$$
$$N_2(g) + 8H^+(aq) + 6e^- \rightarrow 2NH_4^+(aq)$$

6. d.

The only thing needed is two electrons on the right to balance the charges.

$$Sn^{2+}(aq) \rightarrow Sn^{4+}(aq) + 2e^-$$

6. e.

$$ClO_3^-(aq) + 6H^+(aq) \rightarrow Cl^-(aq) + 3H_2O(l)$$
$$ClO_3^-(aq) + 6H^+(aq) + 6e^- \rightarrow Cl^-(aq) + 3H_2O(l)$$

6. f.

$$OH^-(aq) + H_2O(l) \rightarrow O_2(g) + 3H^+(aq)$$
$$OH^-(aq) + H_2O(l) + 3OH^-(aq) \rightarrow O_2(g) + 3H^+(aq) + 3OH^-(aq)$$
$$H_2O(l) + 4OH^-(aq) \rightarrow O_2(g) + 2H_2O(l)$$
$$4OH^-(aq) \rightarrow O_2(g) + H_2O(l)$$
$$4OH^-(aq) \rightarrow O_2(g) + H_2O(l) + 4e^-$$

---

6. g.

$$H_2SO_3(aq) + H_2O(l) \rightarrow SO_4^{2-}(aq) + 4H^+(aq)$$
$$H_2SO_3(aq) + H_2O(l) \rightarrow SO_4^{2-}(aq) + 4H^+(aq) + 2e^-$$

---

6. h.

$$O_2(g) + 4H^+(aq) \rightarrow H_2O(l) + H_2O(l)$$
$$O_2(g) + 4H^+(aq) + 4OH^-(aq) \rightarrow 2H_2O(l) + 4OH^-(aq)$$
$$O_2(g) + 4H_2O \rightarrow 2H_2O(l) + 4OH^-(aq)$$
$$O_2(g) + 2H_2O + 4e^- \rightarrow 4OH^-(aq)$$

7.

In the following solutions, the first set of equations isolates the species involved in the oxidation and reduction. The second set of equations derives the complete pair of half-reactions. The third set of equations shows the addition of the half-reactions, cancellations, and addition of ions to form complete compounds.

7. a.

$$FeCl_3(aq) + H_2S(aq) \rightarrow FeCl_2(aq) + HCl(aq) + S(s)$$
$$Fe^{3+}(aq) + 3Cl^-(aq) + 2H^+(aq) + S^{2-}(aq) \rightarrow Fe^{2+}(aq) + 2Cl^-(aq) + H^+(aq) + Cl^-(aq) + S(s)$$
$$Fe^{3+}(aq) + S^{2-}(aq) \rightarrow Fe^{2+}(aq) + S(s)$$

$$2Fe^{3+}(aq) + 2e^- \rightarrow 2Fe^{2+}(aq)$$
$$S^{2-}(aq) \rightarrow S(s) + 2e^-$$

$$2Fe^{3+}(aq) + S^{2-}(aq) \rightarrow 2Fe^{2+}(aq) + S(s)$$
$$2Fe^{3+}(aq) + S^{2-}(aq) + 6Cl^-(aq) + 2H^+(aq) \rightarrow 2Fe^{2+}(aq) + S(s) + 6Cl^-(aq) + 2H^+(aq)$$
$$2FeCl_3(aq) + H_2S(aq) \rightarrow 2FeCl_2(aq) + S(s) + 2HCl(aq)$$

---

7. b.

$$I_2(s) + HClO(aq) \rightarrow HIO_3(aq) + HCl(aq)$$
$$I_2(s) + H^+(aq) + ClO^-(aq) \rightarrow H^+(aq) + IO_3^-(aq) + H^+(aq) + Cl^-(aq)$$
$$I_2(s) + ClO^-(aq) \rightarrow IO_3^-(aq) + Cl^-(aq)$$

$$I_2(s) \rightarrow IO_3^-(aq)$$

$$I_2(s) \rightarrow 2IO_3^-(aq)$$

$$I_2(s) + 6H_2O(l) \rightarrow 2IO_3^-(aq) + 12H^+(aq)$$

$$I_2(s) + 6H_2O(l) \rightarrow 2IO_3^-(aq) + 12H^+(aq) + 10e^-$$

$$ClO^-(aq) \rightarrow Cl^-(aq)$$

$$ClO^-(aq) + 2H^+(aq) \rightarrow Cl^-(aq) + H_2O(l)$$

$$ClO^-(aq) + 2H^+(aq) + 2e^- \rightarrow Cl^-(aq) + H_2O(l)$$

$$I_2(s) + 6H_2O(l) \rightarrow 2IO_3^-(aq) + 12H^+(aq) + 10e^-$$

$$5ClO^-(aq) + 10H^+(aq) + 10e^- \rightarrow 5Cl^-(aq) + 5H_2O(l)$$

$$I_2(s) + 6H_2O(l) + 5ClO^-(aq) + 10H^+(aq) \rightarrow 2IO_3^-(aq) + 12H^+(aq) + 5Cl^-(aq) + 5H_2O(l)$$

$$I_2(s) + H_2O(l) + 5HClO(aq) + 5H^+(aq) \rightarrow 2HIO_3(aq) + 5H^+(aq) + 5HCl(aq)$$

$$I_2(s) + H_2O(l) + 5HClO(aq) \rightarrow 2HIO_3(aq) + 5HCl(aq)$$

---

7. c.

$$KMnO_4(aq) + CH_3OH(aq) \rightarrow Mn(OH)_2(aq) + HCOOH(aq) + KOH(aq)$$

$$K^+(aq) + MnO_4^-(aq) + CH_3OH(aq) \rightarrow$$
$$Mn^{2+}(aq) + 2OH^-(aq) + H^+(aq) + COOH^-(aq) + K^+(aq) + OH^-(aq)$$

$$MnO_4^-(aq) + CH_3OH(aq) \rightarrow Mn^{2+}(aq) + COOH^-(aq)$$

$$MnO_4^-(aq) \rightarrow Mn^{2+}(aq)$$

$$MnO_4^-(aq) + 8H^+(aq) \rightarrow Mn^{2+}(aq) + 4H_2O(l)$$

$$MnO_4^-(aq) + 8H^+(aq) + 5e^- \rightarrow Mn^{2+}(aq) + 4H_2O(l)$$

$$CH_3OH(aq) \rightarrow COOH^-(aq)$$

$$CH_3OH(aq) + H_2O(l) \rightarrow COOH^-(aq) + 5H^+(aq)$$

$$CH_3OH(aq) + H_2O(l) \rightarrow COOH^-(aq) + 5H^+(aq) + 4e^-$$

$4MnO_4^-(aq) + 32H^+(aq) + 20e^- \rightarrow 4Mn^{2+}(aq) + 16H_2O(l)$

$5CH_3OH(aq) + 5H_2O(l) \rightarrow 5COOH^-(aq) + 25H^+(aq) + 20e^-$

$4MnO_4^-(aq) + 32H^+(aq) + 5CH_3OH(aq) + 5H_2O(l) \rightarrow$
$$4Mn^{2+}(aq) + 16H_2O(l) + 5COOH^-(aq) + 25H^+(aq)$$

$4MnO_4^-(aq) + 12H^+(aq) + 5CH_3OH(aq) \rightarrow 4Mn^{2+}(aq) + 11H_2O(l) + 5HCOOH(aq)$

$4MnO_4^-(aq) + 12H^+(aq) + 5CH_3OH(aq) + 4K^+(aq) + 12OH^-(aq) \rightarrow$
$$4Mn^{2+}(aq) + 11H_2O(l) + 5HCOOH(aq) + 4K^+(aq) + 12OH^-(aq)$$

$4KMnO_4(aq) + 5CH_3OH(aq) + 12H_2O \rightarrow$
$$4Mn(OH)_2(aq) + 11H_2O(l) + 5HCOOH(aq) + 4KOH(aq)$$

$4KMnO_4(aq) + 5CH_3OH(aq) + H_2O \rightarrow 4Mn(OH)_2(aq) + 5HCOOH(aq) + 4KOH(aq)$

---

7. d.

$H_2O_2(aq) + ClO_2(aq) \rightarrow HClO_2(aq) + O_2(g)$

$H_2O_2(aq) + ClO_2(aq) \rightarrow H^+(aq) + ClO_2^-(aq) + O_2(g)$

$H_2O_2(aq) + ClO_2(aq) \rightarrow ClO_2^-(aq) + O_2(g)$

$H_2O_2(aq) \rightarrow O_2(g)$

$H_2O_2(aq) + 2OH^-(aq) \rightarrow O_2(g) + 2H^+(aq) + 2OH^-(aq)$

$H_2O_2(aq) + 2OH^-(aq) \rightarrow O_2(g) + 2H^+(aq) + 2OH^-(aq) + 2e^-$

$ClO_2(aq) \rightarrow ClO_2^-(aq)$

$ClO_2(aq) + e^- \rightarrow ClO_2^-(aq)$

$H_2O_2(aq) + 2OH^-(aq) \rightarrow O_2(g) + 2H^+(aq) + 2OH^-(aq) + 2e^-$

$2ClO_2(aq) + 2e^- \rightarrow 2ClO_2^-(aq)$

$H_2O_2(aq) + 2OH^-(aq) + 2ClO_2(aq) \rightarrow$
$$O_2(g) + 2H^+(aq) + 2OH^-(aq) + 2ClO_2^-(aq)$$

$H_2O_2(aq) + 2OH^-(aq) + 2ClO_2(aq) \rightarrow O_2(g) + 2H_2O(l) + 2ClO_2^-(aq)$

$H_2O_2(aq) + 2OH^-(aq) + 2ClO_2(aq) + 2H^+(aq) \rightarrow$
$$O_2(g) + 2H_2O(l) + 2ClO_2^-(aq) + 2H^+(aq)$$

$H_2O_2(aq) + 2H_2O(l) + 2ClO_2(aq) \rightarrow O_2(g) + 2H_2O(l) + 2HClO_2(aq)$

$H_2O_2(aq) + 2ClO_2(aq) \rightarrow O_2(g) + 2HClO_2(aq)$

---

7. e.

$$Al(s) + LiCl(aq) + LiNO_2(aq) \rightarrow LiAlO_2(aq) + NH_4Cl(aq)$$

$$Al(s) + Li^+(aq) + Cl^-(aq) + Li^+(aq) + NO_2^-(aq) \rightarrow Li^+(aq) + AlO_2^-(aq) + NH_4^+(aq) + Cl^-(aq)$$

$$Al(s) + NO_2^-(aq) \rightarrow AlO_2^-(aq) + NH_4^+(aq)$$

$$Al(s) \rightarrow AlO_2^-(aq)$$

$$Al(s) + 2H_2O(l) \rightarrow AlO_2^-(aq) + 4H^+(aq)$$

$$Al(s) + 2H_2O(l) + 4OH^-(aq) \rightarrow AlO_2^-(aq) + 4H^+(aq) + 4OH^-(aq)$$

$$Al(s) + 2H_2O(l) + 4OH^-(aq) \rightarrow AlO_2^-(aq) + 4H^+(aq) + 4OH^-(aq) + 3e^-$$

$$NO_2^-(aq) \rightarrow NH_4^+(aq)$$

$$NO_2^-(aq) + 8H^+(aq) \rightarrow NH_4^+(aq) + 2H_2O(l)$$

$$NO_2^-(aq) + 8H^+(aq) + 8OH^-(aq) \rightarrow NH_4^+(aq) + 2H_2O(l) + 8OH^-(aq)$$

$$NO_2^-(aq) + 8H^+(aq) + 8OH^-(aq) + 6e^- \rightarrow NH_4^+(aq) + 2H_2O(l) + 8OH^-(aq)$$

$$2Al(s) + 4H_2O(l) + 8OH^-(aq) \rightarrow 2AlO_2^-(aq) + 8H^+(aq) + 8OH^-(aq) + 6e^-$$

$$NO_2^-(aq) + 8H^+(aq) + 8OH^-(aq) + 6e^- \rightarrow NH_4^+(aq) + 2H_2O(l) + 8OH^-(aq)$$

$$2Al(s) + 4H_2O(l) + 8OH^-(aq) + NO_2^-(aq) + 8H^+(aq) + 8OH^-(aq) \rightarrow$$

$$2AlO_2^-(aq) + 8H^+(aq) + 8OH^-(aq) + NH_4^+(aq) + 2H_2O(l) + 8OH^-(aq)$$

$$2Al(s) + 12H_2O(l) + 8OH^-(aq) + NO_2^-(aq) \rightarrow$$

$$2AlO_2^-(aq) + NH_4^+(aq) + 10H_2O(l) + 8OH^-(aq)$$

$$2Al(s) + 2H_2O(l) + 8OH^-(aq) + NO_2^-(aq) \rightarrow$$

$$2AlO_2^-(aq) + NH_4^+(aq) + 8OH^-(aq)$$

$$2Al(s) + 2H_2O(l) + 8OH^-(aq) + NO_2^-(aq) + 2Li^+(aq) + Cl^-(aq) \rightarrow$$

$$2AlO_2^-(aq) + NH_4^+(aq) + 8OH^-(aq) + 2Li^+(aq) + Cl^-(aq)$$

$$2Al(s) + 2H_2O(l) + LiNO_2(aq) + LiCl(aq) \rightarrow 2LiAlO_2(aq) + NH_4Cl(aq)$$

---

7. f.

$$HCl(aq) + HMnO_4(aq) \rightarrow Cl_2(g) + MnCl_2(aq)$$

$$H^+(aq) + Cl^-(aq) + H^+(aq) + MnO_4^-(aq) \rightarrow Cl_2(g) + Mn^{2+}(aq) + 2Cl^-(aq)$$

$$Cl^-(aq) + MnO_4^-(aq) \rightarrow Cl_2(g) + Mn^{2+}(aq) + 2Cl^-(aq)$$

$$Cl^-(aq) \rightarrow Cl_2(g) + 2Cl^-(aq)$$

$$4Cl^-(aq) \rightarrow Cl_2(g) + 2Cl^-(aq) + 2e^-$$

$$MnO_4^-(aq) \rightarrow Mn^{2+}(aq)$$

$$MnO_4^-(aq) + 8H^+(aq) \rightarrow Mn^{2+}(aq) + 4H_2O(l)$$

$$MnO_4^-(aq) + 8H^+(aq) + 5e^- \rightarrow Mn^{2+}(aq) + 4H_2O(l)$$

$$20Cl^-(aq) \rightarrow 5Cl_2(g) + 10Cl^-(aq) + 10e^-$$
$$2MnO_4^-(aq) + 16H^+(aq) + 10e^- \rightarrow 2Mn^{2+}(aq) + 8H_2O(l)$$
$$20Cl^-(aq) + 2MnO_4^-(aq) + 16H^+(aq) \rightarrow 5Cl_2(g) + 10Cl^-(aq) + 2Mn^{2+}(aq) + 8H_2O(l)$$
$$14HCl(aq) + 6Cl^-(aq) + 2HMnO_4(aq) \rightarrow 5Cl_2(g) + 6Cl^-(aq) + 2MnCl_2(aq) + 8H_2O(l)$$
$$14HCl(aq) + 2HMnO_4(aq) \rightarrow 5Cl_2(g) + 2MnCl_2(aq) + 8H_2O(l)$$

---

7. g.

$$S(s) + HNO_3(aq) \rightarrow H_2SO_3(aq) + N_2O(g)$$
$$S(s) + H^+(aq) + NO_3^-(aq) \rightarrow 2H^+(aq) + SO_3^{2-}(aq) + N_2O(g)$$
$$S(s) + NO_3^-(aq) \rightarrow SO_3^{2-}(aq) + N_2O(g)$$

$$S(s) \rightarrow SO_3^{2-}(aq)$$
$$S(s) + 3H_2O(l) \rightarrow SO_3^{2-}(aq) + 6H^+(aq)$$
$$S(s) + 3H_2O(l) \rightarrow SO_3^{2-}(aq) + 6H^+(aq) + 4e^-$$
$$NO_3^-(aq) \rightarrow N_2O(g)$$
$$2NO_3^-(aq) \rightarrow N_2O(g)$$
$$2NO_3^-(aq) + 10H^+(aq) \rightarrow N_2O(g) + 5H_2O(l)$$
$$2NO_3^-(aq) + 10H^+(aq) + 8e^- \rightarrow N_2O(g) + 5H_2O(l)$$

$$2S(s) + 6H_2O(l) \rightarrow 2SO_3^{2-}(aq) + 12H^+(aq) + 8e^-$$
$$2NO_3^-(aq) + 10H^+(aq) + 8e^- \rightarrow N_2O(g) + 5H_2O(l)$$
$$2S(s) + 6H_2O(l) + 2NO_3^-(aq) + 10H^+(aq) \rightarrow 2SO_3^{2-}(aq) + 12H^+(aq) + N_2O(g) + 5H_2O(l)$$
$$2S(s) + H_2O(l) + 2HNO_3(aq) + 8H^+(aq) \rightarrow 2H_2SO_3(aq) + 8H^+(aq) + N_2O(g)$$
$$2S(s) + H_2O(l) + 2HNO_3(aq) \rightarrow 2H_2SO_3(aq) + N_2O(g)$$

---

7. h.

$$HBr(aq) + HMnO_4(aq) \rightarrow HBrO_3(aq) + MnO_2(s)$$
$$H^+(aq) + Br^-(aq) + H^+(aq) + MnO_4^-(aq) \rightarrow H^+(aq) + BrO_3^-(aq) + MnO_2(s)$$
$$Br^-(aq) + MnO_4^-(aq) \rightarrow BrO_3^-(aq) + MnO_2(s)$$

$$Br^-(aq) \rightarrow BrO_3^-(aq)$$
$$Br^-(aq) + 3H_2O(l) \rightarrow BrO_3^-(aq) + 6H^+(aq)$$
$$Br^-(aq) + 3H_2O(l) \rightarrow BrO_3^-(aq) + 6H^+(aq) + 6e^-$$
$$MnO_4^-(aq) \rightarrow MnO_2(s)$$
$$MnO_4^-(aq) + 4H^+(aq) \rightarrow MnO_2(s) + 2H_2O(l)$$
$$MnO_4^-(aq) + 4H^+(aq) + 3e^- \rightarrow MnO_2(s) + 2H_2O(l)$$

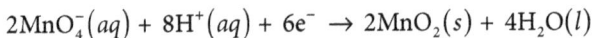

$$Br^-(aq) + 3H_2O(l) \rightarrow BrO_3^-(aq) + 6H^+(aq) + 6e^-$$
$$2MnO_4^-(aq) + 8H^+(aq) + 6e^- \rightarrow 2MnO_2(s) + 4H_2O(l)$$
$$HBr(aq) + 2HMnO_4(aq) + 5H^+(aq) \rightarrow HBrO_3(aq) + 5H^+(aq) + 2MnO_2(s) + H_2O(l)$$
$$HBr(aq) + 2HMnO_4(aq) \rightarrow HBrO_3(aq) + 2MnO_2(s) + H_2O(l)$$

9. c.

$$65 \text{ g Fe}_2O_3 \cdot \frac{\text{mol}}{159.69 \text{ g}} = 0.407 \text{ mol Fe}_2O_3$$

$$0.407 \text{ mol Fe}_2O_3 \cdot \frac{2 \text{ mol Al}}{1 \text{ mol Fe}_2O_3} = 0.814 \text{ mol Al}$$

$$0.814 \text{ mol Al} \cdot \frac{26.98 \text{ g}}{\text{mol}} = 22 \text{ g Al}$$

9. d.

$$65 \text{ g Fe}_2O_3 \cdot \frac{\text{mol}}{159.69 \text{ g}} = 0.407 \text{ mol Fe}_2O_3$$

$$0.407 \text{ mol Fe}_2O_3 \cdot \frac{2 \text{ mol Fe}}{1 \text{ mol Fe}_2O_3} = 0.814 \text{ mol Fe}$$

$$0.814 \text{ mol Fe} \cdot \frac{55.85 \text{ g}}{\text{mol}} = 45 \text{ g Fe}$$

19. a.

$$E^\circ_{cell} = E_{cathode} - E_{anode} = 0.7996 \text{ V} - (-0.1375 \text{ V}) = 0.9371 \text{ V}$$

19. b.

$$E^\circ_{cell} = E_{cathode} - E_{anode} = 0.771 \text{ V} - (-0.28 \text{ V}) = 1.05 \text{ V}$$

19. c.

$$E^\circ_{cell} = E_{cathode} - E_{anode} = 1.066 \text{ V} - (-2.71 \text{ V}) = 3.78 \text{ V}$$

19. d.

$$E^\circ_{cell} = E_{cathode} - E_{anode} = 1.066 \text{ V} - (0.5355 \text{ V}) = 0.530 \text{ V}$$

24. a.

$$1.27 \text{ g NO}_2 \cdot \frac{\text{mol}}{46.01 \text{ g}} = 0.02761 \text{ mol NO}_2$$

$$0.02761 \text{ mol NO}_2 \cdot \frac{1 \text{ mol N}_2\text{O}_5}{2 \text{ mol NO}_2} = 0.01380 \text{ mol N}_2\text{O}_5$$

$$0.01380 \text{ mol N}_2\text{O}_5 \cdot \frac{108.01 \text{ g}}{\text{mol}} = 1.49 \text{ g}$$

$$\frac{1.20 \text{ g}}{1.49 \text{ g}} \cdot 100\% = 80.5\%$$

24. b.

$$4.0 \text{ mol O}_3 \cdot \frac{1 \text{ mol O}_2}{1 \text{ mol O}_3} = 4.0 \text{ mol O}_2$$

$$4.0 \text{ mol O}_2 \cdot \frac{32.00 \text{ g}}{\text{mol}} = 128 \text{ g}$$

$$\frac{x}{128 \text{ g}} = 0.760 \rightarrow x = 0.760 \cdot 128 \text{ g} = 97 \text{ g}$$

24. c.

$$4.304 \text{ g NO}_2 \cdot \frac{\text{mol}}{46.01 \text{ g}} = 0.09354 \text{ mol NO}_2$$

$$0.09354 \text{ mol NO}_2 \cdot \frac{1 \text{ mol O}_3}{2 \text{ mol NO}_2} = 0.04677 \text{ mol O}_3$$

$$0.04677 \text{ mol O}_3 \cdot \frac{48.00 \text{ g}}{\text{mol}} = 2.245 \text{ g O}_3$$

More than this much O_3 is available, so NO_2 is the limiting reactant.

25.

$P_1 = 62.4$ atm, gauge

$T_1 = 22.00°C \rightarrow 22.00°C + 273.15 = 295.15$ K

$T_2 = 35.00°C \rightarrow 35.00°C + 273.15 = 308.15$ K

$P_{1,\text{abs}} = P_{\text{gauge}} + P_{\text{atm}} = 62.4 \text{ atm} + 1.0 \text{ atm} = 63.4 \text{ atm}$

$$PV = nRT \rightarrow \frac{P}{T} = \frac{nR}{V} = \text{constant}$$

$$\frac{P_1}{T_1} = \frac{P_2}{T_2} \rightarrow P_2 = P_1 \cdot \frac{T_2}{T_1} = 63.4 \text{ atm} \cdot \frac{308.15 \text{ K}}{295.15 \text{ K}} = 66.2 \text{ atm}$$

$P_{2,\text{gauge}} = P_{2,\text{abs}} - P_{\text{atm}} = 66.2 \text{ atm} - 1.0 \text{ atm} = 65.2 \text{ atm, gauge}$

26.

$$T = 16.0°C \rightarrow 16.0°C + 273.15 = 289.15 \text{ K}$$

$$P_T = 102,360 \text{ Pa} \cdot \frac{1 \text{ kPa}}{100,000 \text{ Pa}} = 102.360 \text{ kPa}$$

$$V = 275 \text{ mL} \cdot \frac{1 \text{ L}}{1000 \text{ mL}} = 0.275 \text{ L}$$

At these conditions, the vapor pressure of water is 2.0647 kPa

$$P_T = P_{ethane} + P_{water}$$

$$P_{ethane} = P_T - P_{water} = 102.360 \text{ kPa} - 2.0647 \text{ kPa} = 100.295 \text{ kPa}$$

$$R = 8.314 \frac{\text{L} \cdot \text{kPa}}{\text{mol} \cdot \text{K}}$$

$$PV = nRT \rightarrow n = \frac{PV}{RT} = \frac{100.295 \text{ kPa} \cdot 0.275 \text{ L}}{8.314 \dfrac{\text{L} \cdot \text{kPa}}{\text{mol} \cdot \text{K}} \cdot 289.15 \text{ K}} = 0.0115 \text{ mol}$$

27.

$$200 \cdot 325 \text{ mg} \cdot \frac{1 \text{ g}}{1000 \text{ mg}} = 65.0 \text{ g C}_9\text{H}_8\text{O}_4$$

$$65.0 \text{ g C}_9\text{H}_8\text{O}_4 \cdot \frac{\text{mol}}{180.2 \text{ g}} = 0.3608 \text{ mol C}_9\text{H}_8\text{O}_4$$

$$0.3608 \text{ mol C}_9\text{H}_8\text{O}_4 \cdot \frac{6.022 \times 10^{23} \text{ particles}}{\text{mol}} = 2.173 \times 10^{23} \text{ particles}$$

Each particle is a molecule containing 9 carbon atoms.

$$2.173 \times 10^{23} \cdot 9 = 1.96 \times 10^{24} \text{ C atoms}$$

28.

$$CH_3CH_2OH + 3O_2 \rightarrow 3H_2O + 2CO_2$$

$$1.00 \text{ L H}_2\text{O} \cdot \frac{1000 \text{ mL}}{1 \text{ L}} \cdot \frac{998 \text{ g}}{\text{mL}} = 998 \text{ g H}_2\text{O}$$

$$998 \text{ g H}_2\text{O} \cdot \frac{\text{mol}}{18.02 \text{ g}} = 55.38 \text{ mol H}_2\text{O}$$

$$55.38 \text{ mol H}_2\text{O} \cdot \frac{1 \text{ mol CH}_3\text{CH}_2\text{OH}}{3 \text{ mol H}_2\text{O}} = 18.46 \text{ mol CH}_3\text{CH}_2\text{OH}$$

$$18.46 \text{ mol CH}_3\text{CH}_2\text{OH} \cdot \frac{46.07 \text{ g}}{\text{mol}} = 850.5 \text{ g CH}_3\text{CH}_2\text{OH}$$

$$850.5 \text{ g CH}_3\text{CH}_2\text{OH} \cdot \frac{\text{cm}^3}{0.789 \text{ g}} = 1080 \text{ cm}^3 = 1080 \text{ mL} \cdot \frac{1 \text{ L}}{1000 \text{ mL}} = 1.08 \text{ L}$$

---

www.ingramcontent.com/pod-product-compliance
Lightning Source LLC
Chambersburg PA
CBHW081550220326
41598CB00036B/6629

* 9 7 8 0 9 9 8 1 6 9 9 0 3 *